Science and Fiction

Science and Fiction – A Springer Series

This collection of entertaining and thought-provoking books will appeal equally to science buffs, scientists and science-fiction fans. It was born out of the recognition that scientific discovery and the creation of plausible fictional scenarios are often two sides of the same coin. Each relies on an understanding of the way the world works, coupled with the imaginative ability to invent new or alternative explanations—and even other worlds. Authored by practicing scientists as well as writers of hard science fiction, these books explore and exploit the borderlands between accepted science and its fictional counterpart. Uncovering mutual influences, promoting fruitful interaction, narrating and analyzing fictional scenarios, together they serve as a reaction vessel for inspired new ideas in science, technology, and beyond.

Whether fiction, fact, or forever undecidable: the Springer Series "Science and Fiction" intends to go where no one has gone before!

Its largely non-technical books take several different approaches. Journey with their authors as they

- Indulge in science speculation – describing intriguing, plausible yet unproven ideas;
- Exploit science fiction for educational purposes and as a means of promoting critical thinking;
- Explore the interplay of science and science fiction – throughout the history of the genre and looking ahead;
- Delve into related topics including, but not limited to: science as a creative process, the limits of science, interplay of literature and knowledge;
- Tell fictional short stories built around well-defined scientific ideas, with a supplement summarizing the science underlying the plot.

Readers can look forward to a broad range of topics, as intriguing as they are important. Here just a few by way of illustration:

- Time travel, superluminal travel, wormholes, teleportation
- Extraterrestrial intelligence and alien civilizations
- Artificial intelligence, planetary brains, the universe as a computer, simulated worlds
- Non-anthropocentric viewpoints
- Synthetic biology, genetic engineering, developing nanotechnologies
- Eco/infrastructure/meteorite-impact disaster scenarios
- Future scenarios, transhumanism, posthumanism, intelligence explosion
- Virtual worlds, cyberspace dramas
- Consciousness and mind manipulation

More information about this series at http://www.springer.com/series/11657

Michael Carroll

Plato's Labyrinth

Dinosaurs, Ancient Greeks, and Time Travelers

 Springer

Michael Carroll
Parker, CO, USA

ISSN 2197-1188   ISSN 2197-1196   (electronic)
Science and Fiction
ISBN 978-3-030-91708-1   ISBN 978-3-030-91709-8   (eBook)
https://doi.org/10.1007/978-3-030-91709-8

Cover image © FlashMovie/Getty Images/iStock

This Springer imprint is published by the registered company Springer Nature Switzerland AG
The registered company address is: Gewerbestrasse 11, 6330 Cham, Switzerland

I'd like to dedicate this little story to my fellow traveler and decades-long friend, Bill Gerrish.

"Time's wheel runs back or stops: potter and clay endure."
                                                    -Robert Browning

"Anyone who is not shocked by quantum theory has not understood it."
                                                    -Niels Bohr

"Time is what keeps everything from happening at once."
                                                    -Ray Cummings

Acknowledgements

First, and always, my loving thanks goes to my wife, who helps me see the holes in my plots and gives me energy on those off days when I want to give up. My fellow artist Marilynn Flynn is always there for critique and encouragement. As with other projects, Hannah Kaufman, Springer editor extraordinaire, husbanded this project through the many hoops through which a book must leap to see the light of day. My gratitude to both her and Mark Alpert, the Science and Fiction series editor. Special commendation goes to my quantum physicist buddies Kara Szathmary and Mark Garlick, who are not only fine scientists but also excellent artists (some people get all the talent!). Physicist Ronald Mallett put up with silly questions to help me portray a reasonable time travel scenario.

Speaking of physicists, talented physicist/author Randy Ingermanson, an award-winning writer, provided invaluable (that means *really* valuable) feedback and wisdom on my manuscript. Brent Breithaupt, Regional Paleontologist at the BLM, offered valuable background on all those prehistoric beasts, as well as the workings of the dinosaur-digging profession. Andy and Vicki Chaikin provided moral support and New York bus line details. Thanks to Betty, Alan, and all my friends at Doug's Diner for pumping me full of coffee at just the right times, and my Life Group buddy Ken Anderson for advice on firearms and holsters. Jan Siegel, Rare Book Librarian of Columbia University's Rare Book and Manuscript Library, opened the world of ancient documents to me, helping me form a more realistic view of how curators would handle those sketches by Benjamin Waterhouse Hawkins. Tamar McKee of the Stephen H. Hart Research Center at History Colorado helped me obtain some great materials related to document preservation.

A special shoutout goes to Dr. Kristin Heineman, Associate Teaching Professor in Colorado State University's history department, for her valuable insights and advice on the Neopalatial Minoans. I wish you had been teaching back when I was at CSU! Denver University's Dr. Victor Castellani offered his linguistic expertise to the story. My own brother-in-law, Russ Wiggam, gave me road directions for Como Bluff and Medicine Bow, lending an air of "hey, this-author-knows-what-he's-talking-about" to the text, even though when it comes to Wyoming, I don't. My favorite travel agent, Terri Engel, checked my details for the Bahamas (since I haven't made it there yet). My writing comrades, Michelle McCorkle and Carol Eaton, along with the Mile High Scribes, provided insights both grammatical and motivational, to flesh out my characters and plot. They also helped me to keep my head on straight.

A good story takes many experiences, and a book takes a team. Thank you all.

Contents

Part II   The Science Behind the Story

Part I

Novel

1

Bradley Glenn
The Present, New York City

Bradley Glenn had yet to down his first cup of coffee when the phone rang. It was the museum's landline, and the only outlet rested across the room on the boss's desk. The Director of Paleontology, Ivan Trask, held up the handset in his beefy paw.

"Brad, you want to take this?"

"Is that a real question?"

A grin split the Director's wide face. "It's the foreman of some construction site across the Park. Bones. Sounds very important. Could be the find of the century."

The *find of the century* was code for *yet another looney who thinks he's found a dinosaur in the backyard.*

Brad sauntered across the room, weaving his way between piles of dusty file folders, snaking computer cables, and shelves of bone fragments. He gazed ruefully back at his cooling mug of Columbian Dark. Emblazoned across its face were the words *American Museum of Nature and Science.* He grabbed the phone from his boss in mock fury.

"Paleontology. Dr. Glenn here."

"Hello, Dr. Glenn. This is Neil Battaglia. I'm on the crew who's renovating the zoo. The Central Park Zoo?"

"Yes, I've dealt with the traffic. Seems like you all are doing some major things down there."

Battaglia let out a primal grunt. "We're workin' on the Birds and Penguins exhibit. Water. It's the worst. Ask any architect, any bricklayer, any electrician. Water's your worst enemy. You drain everything and then fix and then you

© The Author(s), under exclusive license to Springer Nature Switzerland AG 2021
M. Carroll, *Plato's Labyrinth*, Science and Fiction,
https://doi.org/10.1007/978-3-030-91709-8_1

don't find the real problems until you fill it all up again. It's a real pile of sh—a real hassle."

"Sounds like." Bradley tried to sound sympathetic. "I understand you think you found some dinosaur bones."

"Not bones, exactly. Hold on." The man's voice cranked up an octave away from the phone. "Not there, guys, over there. No…there. Don't you guys read? And brace that thing. We don't want somebody gettin' hurt again." He dropped his tone to a confidential level. "Assholes. Sorry mister— *Doctor*—Glenn."

"You were saying?" Bradley encouraged.

"So, yeah, not dinosaurs, exactly. Not bones, but the real thing."

He had Bradley's attention. "The real thing?"

"Put it this way: one of 'em's staring at me right now."

*   *   *

Bradley wasn't sure what Ivan Trask would make of him dashing across Central Park to see a "real" dinosaur, but he knew his boss was all in when Ivan said, "This I gotta see."

In the workstation next to Bradley, Delaney Delgado peered over her monitor, her bleached hair forming a wispy halo around her mahogany face. Around the pencil in her mouth, she said, "Sounds like another Laurel and Hardy escapade." The two men grinned.

Ivan stood. "Delaney, can you hold down the fort?"

She tossed the pencil down onto the desk. "This smacks of sexism."

"What this smacks of is you're the newest one here, and you got to go on the Colorado Springs trip when we left Brad behind."

She punched her glasses up the bridge of her nose, eyes blazing beneath her faux-blonde bangs. "You two always run off on the fun boondoggles and leave the hardware store runs for me. We got nothin' going on here until Exhibits sends us the order for 'Hadrosaurs of Hudson River.' You need me to come along. You two hoodlums need stability. If I didn't know you better, I'd suspect it was all because I'm a member of a gender not yours."

"It could be because you are a member of Looney Tunes," Ivan retorted. "Do you think you could present a professional front to our public?"

She grabbed a Triceratops hat from her desk, stuffed it behind some files, and sat up straight. As Triceratops hats went, this one was fairly conservative, but she was on company time. "Of course boss."

"It does seem reasonable," Bradley said.

Ivan nodded at Delaney. "Three heads are better than two. Besides, you're the best preparator this place has had in a zillion years, not that I would ever admit that out loud."

"Might go to her head," Bradley added.

"We may need your expertise for digging out whatever these folks have uncovered."

They could have taken a crosstown bus from Central Park West, but Ivan popped for a taxi, faster and not much more. The walk from the cab to the zoo was bustling, loud, and sunny. The City offered its curious combinations of aromas—diesel fuel and curbside hot dogs, cool concrete and hot glass, cigarette smoke and sidewalk fruit stands.

The guard at the entrance let them through and pointed them toward the penguin enclosure. The once-arctic concrete pond was a mass of yellow construction tape, plastic cones, and big machines chewing up asphalt slabs and sidewalks. At the far side, a man in a hardhat waved enthusiastically and approached. He looked as though he had started his five-o'clock shadow a week ago.

"Neil Battaglia," he barked, jamming a hand toward Ivan. Introductions were made, and Battaglia stretched his arm out behind him. "Welcome to my kingdom. Shall we?"

Delaney grinned at Bradley, who stifled his smile.

At the far end of the excavation, several rusted pipes protruded from a bank of rich soil. One was leaking. Battaglia cast a baleful glance at it and leaned in toward Ivan. "See? Water. Bah."

He led them into the pit. The fresh morning air carried the scent of wet loam and rusted metal. The four of them stood in a row, baffled. Concrete scraps, many with patterns or faded paint, lay scattered across the bank. Bradley kept himself from nudging the closest fragment, a greenish, beachball-sized cement chunk with scales and one large, reptilian eye staring at him.

"Don't make eye contact," Delaney whispered urgently. "They don't eat you unless they see you."

The three paleontologists stood motionless.

Battaglia broke the silence. "'Course they're not *real* dinosaurs. They're statues or somethin'. Like in a museum. Guess you'd know all about that."

Bradley spoke up. "I do indeed. In fact, you've made a very important discovery. Not so much for us paleontologists, but certainly for historians."

Battaglia looked lost.

Bradley straightened. "If I'm right—"

Delaney put in, "And he's almost never right."

Ignoring her, he went on, "I think—not sure, but I think—these may be the long-lost sculptures of Waterhouse Hawkins. And if they are, even broken up like this, they're priceless."

Battaglia shoved a hand under his helmet and scratched. "And who was he?"

"*Benjamin* Waterhouse Hawkins," Delaney added. "British sculptor."

Battaglia let his helmet fall back into place. "What are all his statues doing in here?"

Bradley leaned over and picked up a concrete tube with a point on the end of it. "Well, Mr. Battaglia, as they say, therein lies a tail."

2

Waterhouse Hawkins in Central Park
1871, New York City

Waterhouse Hawkins paced the concrete-powdered floor like one of the caged leopards in the zoo across the greenway, his hands clasped behind the small of his back. "The Crystal Palace was fine, yes. *The Times* called it a masterpiece. But this: *this* would have been my true masterpiece. How could they do this? I thought this country was founded on some degree of democracy. How can a handful of miscreants—"

A young man, scarcely out of his teens, cowered in the corner. "Mr. Hawkins, you can't go around talking like that. If anybody from Tammany Hall—"

Hawkins didn't hear. "The idea. How can they just heave out every one of the Central Park Commissioners? Every one, without so much as a how do you do?"

"If you can't calm yourself you may have a fit or something."

Hawkins tossed the boy a withering glare. He still found the American accent difficult to interpret. "That sounds like something your parents would say."

"I'm twenty. Totally independent. Have my own place and all."

Hawkins pulled a pair of pince-nez spectacles from his pocket. He had been told they were the hottest in Paris fashion, but he had no time for fashion. He simply liked the way they stayed put while he leaned over a drawing table. He polished the glasses on his handkerchief and began his pacing once again. "And all the work outside. This is about more than our little menagerie here. It's, it's everything else. No more Paleozoic Museum, after all of Mr. Olmstead's work."

M. Carroll, *Plato's Labyrinth*, Science and Fiction,
https://doi.org/10.1007/978-3-030-91709-8_2

The young man glanced from side to side, as if assuring himself that no one was listening. "They say Boss Tweed is mighty powerful. Too powerful to fight."

"Well, your William Tweed is certainly no fan of higher education for the common man," Hawkins blustered. "Just look at it, Tommy. All we've done." Hawkins gestured toward the nearest statue, and then quieted. He sauntered across the room, taking it all in as if it were a scene he had just discovered. "You see, Tommy, there is power in what we do here. Potential. Do you know what that word means?"

"Yessir."

"When the scientist discovers something—a new aspect of God's creation—the world is inspired. And inspiration breeds exploration, and more discovery. And then, best of all, when the artist takes that science and brings it to life in sculpture or painting or literature, the common man shares in an even deeper way. Science and knowledge are like that. They come with responsibility, of course. And Boss Tweed does not understand responsibility. He does not understand the power behind science and invention." Hawkins shook his head, looking away from the sculptures. "Where will my beauties go now?"

He stared out the window into the park. The grounds in the distance had been shoveled and sculpted and manicured, prepared for the gardens and foundations and lovely museum to come. But the workmen were gone, vanished as surely as the funds had disappeared. His eye fell upon one of the beasts this side of the window. It was his favorite, his great Iguanodon, bigger even than his famous Megalosaurus back in Sydenham. The concrete façade had hardened nicely. It sat in a stately pose, legs bent like a reclining dog or lizard, tail curled around itself. All 39 glorious feet of it.

Hawkins had listened carefully to Richard Owen, the man from the museum in London, about the arrangement of scales and the form of claws. But Hawkins had already observed so many reptiles that he knew how the scales should go on his giant lizard. They fit uniformly, like the plates on a knight's armor. Transverse rows spiraled around the body naturally, with scales at the joints smaller than scales on the flank or stomach. Hawkins had given a nod to iguanas and crocodiles when he laid larger scales along the backs of the fingers and toes, and along the edge of the jaw. But he was most proud of the skin—of how it draped like curtains from the back over the shoulders. The belly bulged where it rested against the ground, and the muscles of the limbs rippled beneath the skin. The creature strained against gravity with its hulking weight. No one could tell that within the soft folds of flesh lay an iron frame and bulwarks of brick. That was the trick—Hawkins knew.

"Masterpiece," he murmured.

"Beg pardon?" Tommy said.

"There's good reason that the press called them a masterpiece, but they don't truly understand. My Iguanodon is a case in point: four iron pillars, thirty tons—metric tons, mind you—of clay; six hundred bricks. Why, Tom, it's like building a house upon four columns. And we did it! You and I."

"My back remembers every one of your ninety casks of stone."

"And worth it?" Hawkins asked, somewhat tentatively.

"Well worth it, sir. Well worth it. You'll find a home for the lovely beasts. You and the museum. You'll see."

There was a loud thump on the side door, and then another. It was scarcely one in the afternoon. No one came by so near the American luncheon hour. Hawkins gestured to Tom. "See to that, will you, son?"

"Yessir." Tommy took two steps toward the door when it burst open. Four large oafs stepped in, three carrying sledgehammers, one carrying a clipboard. They could have been part of any construction crew in the park these days, but somehow Hawkins didn't think they had their minds on creating any-thing. Destruction was more their forte.

The gentleman without a hammer stepped forward and took off his hat with a flourish. "Mr. Benjamin Hawkins, I presume?"

"That is I," Hawkins said. He detected a slight Irish brogue in the man's speech. "Would you like some tea?"

The man turned to his crew and they let fly a chorus of laughter. He turned back to Hawkins, but the mirth in his expression had faded to something more unsettling. The man now spoke in a poorly done American caricature of a cockney accent.

"Actually, your highness, we've come with a message. You know of who it's from."

Hawkins' inner thoughts corrected the man with a *whom*, but his vocal cords were petrified. The unwanted visitor continued, "Our benefactor doesn't like pasty Englishmen interfering in his business, sayin' things about crooked deals and kickbacks and so forth. Isn't that how you put it? 'Kickbacks', you said. That was a mistake. So if you'll please just vacate the premises, we'll be finished shortly." He jerked his face toward the door.

The bile rose in Hawkins' throat. He could feel the sweat roll down his back, feel their eyes on him, just hoping he would put up a fight, offer an argument. He did not.

"Tommy. Come," he said, guiding the young man out the door.

"Very reasonable," the man chided as he shut the door behind them.

"Should we call the police?" Tommy asked eagerly when they were outside. But Hawkins had already made contact; a cop on the corner was watching him. His eyes were glued to Hawkins; the frown on his face spoke volumes about where his allegiances truly lay.

"No, Tommy, most of the bobbies around here know already. Come."

Hawkins ushered them to a park bench. They sat, and Hawkins eyed the bobby, who at least had the decency to act as if he were ignoring them. Keeping his gaze on the cop, Hawkins leaned toward Tommy and lowered his voice.

"You see, Tommy, there is power in science. From science comes technology, and there is power in technology, too. It has the power to heal or to destroy, to build up or to tear down. Power confuses some people, makes them afraid. Science can be used for good or evil, and it must be treated responsibly. But this…" He gestured back toward their building with a subtle jerk of his chin. "This is so uncalled for. People like that bobby or Boss Tweed are not ready for such advances."

"Or not interested," Tommy said.

"So true, lad. So true."

They could hear the crash of heavy objects, the thud of hammers and the crack of concrete and brick. For good measure, someone tossed a loose brick through the glass windowpane facing them. Hawkins grimaced and moved Tommy further away from the wall.

The door opened, and they could smell the chalk dust and taste the powdered brick on their tongues. The men exited, grinning, covered in white grime. The leader and his troupe set out across the commons, but one of the toadies stepped back and stage-whispered to Hawkins, "Don't you bother so much about dead animals. There are lots of live animals; you can make models of them." He started to walk away, but paused for an aside over his shoulder. "Some chaps will be along soon to clean all this up for you."

The man joined the others and disappeared into the crowds walking the park on the sunny spring day. Tommy was shaking.

"You don't have to go back in, young man. You've done plenty of work for the day. Why don't you head home?"

Tommy wouldn't take his eyes from the open door. "No, I want to see. I want to see what's still…standing."

Hawkins patted the boy on his shoulder and realized that he himself was shaking, too. He squared his shoulders. "Brave young man. Let's see what Boss Tweed has left us."

3

The Lecture
The Present, Washington, D.C.

The lecturer glowed beneath the spotlight, looking like some biblical messenger from the celestial sphere. But she didn't float across the stage. She stomped, gesticulating energetically, emphasizing every other word.

"We're told by some of my colleagues that time is a river, that it flows along in an inexorable stream, unreeling the events that we see as our lives. I ask you, on whose authority? No, time does *not* flow. It simply transpires. And because it has no 'flow' to it—" Here the speaker used air quotes, "—then the concept of a direction of time vanishes. What seems like a forward current, a tide through the quantum universe, is simply the causal parade of events. Eddington's arrow of time is a convenient mirage."

The speaker pushed a button on her remote. Her laptop advanced to the next screen. White letters against a black background.

Time flies like an arrow,
   But fruit flies like an apple
   anonymous

The audience laughed as she continued, "This misguided idea of time as an arrow, as a flowing force, deceives us into thinking that time has forward momentum, and that in turn lulls even the most discerning of us into believing that we may move within that imaginary continuum."

The man sitting in the dark at the back of the hall thought very little about spiritual matters, but the biblical analogy was irresistible. When the light hit the speaker right, her blonde hair glowed like a halo.

© The Author(s), under exclusive license to Springer Nature Switzerland AG 2021   **11**
M. Carroll, *Plato's Labyrinth*, Science and Fiction,
https://doi.org/10.1007/978-3-030-91709-8_3

"Which brings us to…" The angelic speaker changed her computer's slide. "Closed time-like curves. Now this concept is one of the most exaggerated and certainly misinterpreted concepts from Einstein's General Theory of Relativity."

The listener sat up. He was tall and thin, and the tuft of white hair on top of his head showed even in the gloom of the back row. This was what he came for. The speaker—a genius in her field—seemed to be speaking directly to him.

"All of us, and all things, exist in physical space and in what we call time. In our mathematics, closed time-like curves define a sort of pathway, a trajectory of an object that travels through space/time and ends up at the same point in both space and time. The implication is that travel through time is possible if the math lines up just right. But I encourage you to take note that CTCs are mathematical models. They have absolutely no counterpart in physical reality. CTCs have been called upon as 'time machines'." The speaker sent up her air quotes again. "But I can assure you: there are no technological components that we can cobble together to create such a thing."

The observer let out an involuntary *humph*. He glanced around to make sure no one heard.

"It's not a matter of technology catching up," she continued. "No: the fact is that despite decades of research, argument, and speculation, often *wild* speculation, at the heart of CTCs lies the central component of *time*. The simple problem is that the physics community has yet to decide whether time really exists."

The man tried to swallow a chuckle, but he could feel it building. He began to snicker and headed for the door, trying desperately to exit the room before his reaction became gales of laughter. He was only partially successful.

4

Katya Joshi
The Present, Fort Collins, Colorado

Katya Joshi seldom took lunch. Her boss told her she was driven. "In a good way," he had added weakly. After her morning run, she preferred to stay at work, watching the circus around her. She was surrounded by clowns, but—as her boss would have said—the good kind. The kind who hid their genius behind office antics, practical jokes, and freewheeling postulations. The circus was too good to miss. But today she was making up for lost time, leaving half an hour early. ChronoCorp was laid back enough that it really didn't matter, but it did to her. She liked things in their place, at the right time. Ironic that she would be employed here, where the overarching byproduct was to play with established timelines.

Katya stepped out to the front pavement and glanced back at the glass and steel building, one of a handful of structures in Fort Collins that stood over a couple stories high. Few people knew what really took place on those top few floors. ChronoCorp preferred it that way, for now.

It was a beautiful autumn day, and she had plenty of time to walk the three short blocks to Old Town. The snows would come soon enough, and then everyone would be driving everywhere. She had broken her prescription sunglasses a month ago and was filling in with some Mini-Mouse shades her niece had lent her. A quick right on Mountain Avenue took her directly to the brick pavement of Old Town Square. The rough-hewn Lyons Sandstone buildings rose above sidewalk tables crowned by cheery blue umbrellas. Students gathered around a street performer near a faux waterfall. Beyond the historic buildings and the umbrellas and streets of "The Fort" rose the undulating blue wall of the Rocky Mountains. A dusting of snow lay across the foothills, just

© The Author(s), under exclusive license to Springer Nature Switzerland AG 2021
M. Carroll, *Plato's Labyrinth*, Science and Fiction,
https://doi.org/10.1007/978-3-030-91709-8_4

enough to give them some definition. Beyond those low hills rose the bastions of granite: Mount Evans, Bierstadt, Long's Peak.

Her dad was waiting at Goldsmythe's Pub & Grille. These days, his back bent a bit more than it used to, and his salt and pepper hair had faded to a snowy white—a dramatic frame around his caramel face. All this was crowned by an omnipresent white straw hat. His smile lines were bona fide wrinkles now, but he still had that essence of dad. All his friends had become the faded people: hair and skin drained of color, parchment cheeks and hands, thinning and crooked limbs. But not him. His Indian heritage kept his complexion rich, it was his work in the field that kept him young.

As she stepped through the door, a tsunami of aromas washed over her: grilling onions, smoky bacon, a whiff of singed oil.

"Hey old man!" she cried as she slid into the booth.

He grabbed her hands across the table. "How's my baby?"

She pulled her pink plastic sunglasses off and pointed them at him. "Livin' the dream." She reached over and straightened his nametag. "No one can read who you are at this dinosaur conference of yours." Her voice rose to a squeaky soprano. "Well, what kinda crazy name is that, anyway? Ajit? Is that your first name or last name?"

"I meant to take it off. Meetings, you know?" Hope you don't mind." He gestured toward the table, already set with a burger for her and a Reuben for him. He leaned toward her. "I still can't get over it: Fort Collins. Not Silicon Valley, not Urbana Illinois, but Fort Collins? All that technology and—"

"Dad, we talked about this," she shushed over the general bedlam of the pub.

"Oh, I know," he said conspiratorially, lowering his voice. "Proprietary. Mum's the word. Did you see this, by the way?" He handed her his phone. On the screen was a news story about old dinosaur sculptures in Central Park.

"I did. Isn't that crazy? After all this time? Quite the find. You should be so lucky in Como."

"Como Bluff will not keep its secrets from my team. Not for long. We'll be finding lots of beasts much older than those Central Park critters." He tapped the screen; the photo of a face appeared in the scrolling text.

She waved a French fry at the article. "Know that guy?"

"Sure do. Bright young man. He's with the museum. American Museum of Nature and Science."

Katya nodded, her cheeks stuffed with cheeseburger. "New York."

"Yes." He was peering over the top of his glasses with that look she dreaded. Here it came. "He's very sharp. Definitely your type."

"What, AB positive?"

"More likely negative," he grumbled. "I would never suggest you need someone to 'complete you' as a person. You know that. I just thought the two of you could have some laughs, share some science stuff."

"If we were ever in the same state, I'm sure we'd have lots to talk about."

"You actually met. You were young. My awards banquet."

She dragged a French fry across the catsup on her plate, doodling an abstract design. A smile lit her face.

Her father laughed, holding up his hand defensively. "I know what you're thinking, I know. David Sandwell…my mistake. I cry uncle!"

Now they were both laughing, until suddenly, Ajit quieted.

"So you still think you can do it? ChronoCorp's little project?"

She had to be careful not to show her hand. Her father couldn't know that she had already made three trips—three "Q-slips."

She folded her hands in front of her and dropped her chin, drilling him with her eyes and taking on the tone of a patient grade-school principal. "I do. It's just a matter of time."

Ajit pointed a pickle spear at her. "'Matter of time.' Good one!" A flicker of long-suffering played across his face. "I've been thinking about it a lot, of course. My daughter running off into the *Land Before Time* or into *Futurama*. It gives one pause. But I just can't help coming to the conclusion that—"

"That it's impossible? With all that's happened in just the span of your own lifetime? The human genome? The Higgs boson? Moon landings? It's quite a list you can add up."

"I have to admit, it was pretty cool when they invented the catapult," he said to his empty dish. Then he looked up, met her eyes with a weak smile. "No: unless you know something I don't, I'm afraid we're bolted to this moment." He paused and studied her. "Is there something you know that I don't?"

She didn't answer.

He went on, "I suspect you'll find that there is no wandering, except the kind we can do with our minds and imaginations."

"Like Verne and Wells? Their wanderings were well done."

Ajit grinned. "Or like me and my colleagues uncovering treasures on Como Bluff. Or the archeologists unveiling the past to us through their work in…" He searched for an example. "In Greece, digging up the treasures of the Minoans and such."

Kat felt herself stiffen. She hoped he didn't notice. Why would he pick Greece, of all places? "I really shouldn't talk about it, anything specific, because we're not sure of causal variables."

"Long-term ramifications. Ripples in the pond. Certainly, I understand." He mimed zipping his mouth, then looked out the window at the people passing by. "It still worries me. If it all does work out, you know."

"It's being carried out very conservatively. At first we'll limit ourselves to just a few minutes at a time, with solo trips. And Xavier was—Xavier will be the first." She waited, hoping he hadn't noticed her slip.

Ajit shrugged. "Dr. Stengel's the boss. I'm surprised he can fit those long legs of his into your contraption. He looks loosely put together, like his limbs are longer than they should be. And his head sort of bobbles around on top."

"Yeah, he'd be the first to admit it. He does have to get into the pod in a fetal position. But it makes sense for him to go first. He's got the background for the engineering and he knows the history. Todd Tanaka gets to go next, a little further. Xavier is taking things slow until we're sure all the tech works and it's safe and blah-blah-woof-woof." She said the last in a sing-song.

"Don't be cavalier, Kat. The kind of technology you're exploiting comes with responsibility. Xavier is being prudent."

Katya smiled and shook her head. "His words exactly."

"Sounds like a boy's club dynamic, if Xavier and Todd get to go first. Is that frustrating?"

"My colleagues are good guys." She looked down at the table and took in a deep breath. "Xavier wants me in charge if anything goes wrong with the slips."

Ajit didn't reply immediately. He stared into his glass, studying the little icebergs bobbing against the sides. "I suppose that's a compliment to your leadership skills." He looked up again. "You know, though, I've been thinking a lot about those Hawkins statues in Central Park. About your work." He leaned closer. "Maybe science could be put ahead, given a nudge, if someone traveled back in time. Say, to the Triassic, to a major extinction event."

Katya shook her head firmly. "Can't go any farther back than when modern humans were present. We can't even go to New Zealand or Australia or Hawaii until humans made it there—we've done the math. Something about the nature of the time flow."

"What if you went back to visit one of *them*? A paleontologist who was around before anybody invented the word. Maybe not just any bone digger, but somebody who knew lots of scientists in the field. Well connected. And what if you showed this strategically chosen person ideas, just hints, of how dinosaurs really stood and walked? Go back to London's Royal Museum, or Philadelphia with Cope. It could put us way ahead. Twenty years before now, instead of drawing our dinosaurs with scales, we'd already be giving them feathers and wrinkled skin. Think where we'd be now; think of it!"

"I can tell you've been cooped up with a few hundred other paleontologists for too long."

"Yes, yes, and a good conference it's been. But it's not just paleontology. We could accelerate the progress of humanity. Put medicine decades ahead and cure cancer. Perhaps even—" he faltered.

She watched her father sitting in the little booth, so strong and wise and invincible. Her mind went there again, to the way she could hop into bed with Mom and cuddle against the night or the cold or the bad dreams. Her mother always let her, even when she was trying to sleep off the effects of the chemo. She was like that.

Katya hated to see her father cry. It made the world a little more uncertain, made him a little less formidable. He put on a good show for others, but even after all this time, he still let his guard down with her, let her see how much it hurt.

His face had gone red. His eyes brimmed. Katya fought to keep her emotions at bay, for his sake. She felt the gorge rising in her throat.

"All she would have needed—"

Katya stopped him, steadying her own voice. "Don't. Don't go there."

"Just minutes would have made a difference. I should have been there."

"Don't, Dad. Just stop."

He took a moment to compose himself, then went on, "Your time-surfing buddies could speak into the ear of Tsiolkovsky or von Braun and we'd be colonizing Mars by now. Can you imagine?"

She smiled gently, willing herself to move with him, away from the loss of her mother, his wife and life companion. "Some people say that's playing with fire. Change one little thing."

"Like the Bradbury tale," he sniffed. "Butterfly effect."

"Hey, you've read his story! It's practically required reading for our bunch. Guy steps on a butterfly in the Jurassic, and *poof!* Everything is different in his own time. But really, it's more the opposite of the butterfly effect. We think time sort of heals itself. No one can change anything big, anything that would last. It's the universe's safeguard."

"But that's theory, right?"

"Right."

Ajit let out a long gust of breath. "Well, then, I suppose it's for the best."

He seemed resigned, but Katya's mind raced. What would reality look like if Marielle Joshi were still around? What kind of world would this be if her mother had continued her research? The field of ecology would certainly have benefited, as would the planet. And just imagine the ripples if Katya put into action a cascade of events that led to a cure for pancreatic cancer. Tens of thousands would live fully without suffering; ten times that would carry on without the kind of loss she and her father had endured. Her father would finally be free of his guilt—of that moment in time that had shaped his years ever since. Yes, what could be the harm?

"Someone should try," she muttered.

"What was that?" her father asked.

"Just thinking out loud. We need to push on so we can actually do research in the field."

She needed time to process, to evaluate this grand scheme. Cheerily, she squeezed his hand. "Speaking of working in the field…"

Ajit straightened. "Yes, yes, I should go. The group leaves in time for dinner up in Medicine Bow. Love you." He kissed her on one cheek, then the other, and dropped a pile of cash on the table. "Can you take care of the paperwork?"

"Thanks for lunch, Dad. Knock those sauropods dead."

"I'm pretty sure they already are." He popped his white straw hat onto his head and strode from the restaurant, a man on a mission.

Katya was left with one thought, her father's innocent suggestion. *We could accelerate the progress of humanity.*

What kind of universe would that be?

5

Toying with Eternity

Xavier was in earlier than usual. As head of the company, he often exercised a little freedom in what time he showed up at the office, and he wasn't a morning person by his own admission. Today, he got a gold star for promptness. His tassel of white hair was even combed back and plastered to his scalp. Katya suspected that something must be up.

He greeted her at the door with an unintelligible train of syllables. "I've now learned how to say, 'can I have some new clothes?' in ancient Hurrian. I'm working on some of the ancient Anatolian words too, but it's tough going."

Katya glanced up at the towering man, a telephone pole with thick glasses. "Hurrian, huh?"

He nodded. "Related to the ancient Minoan. They think."

"I suppose the whole team should get fluent in these ancient tongues you keep throwing around. Maybe we should have another staff retreat to your little cabin in the woods."

"Poudre Canyon is getting a bit cold for our mountain getaways. We'll have to wait for May or so."

"I suppose," Katya agreed.

Xavier offered a wan smile. "Ironic, all this. Who knew our studies would lead to actual—" He waved his hand around as if pointing to their surroundings would fill in the blank.

"Yep. I remember those carefree days before all this: eighteen hour shifts, sweating over computer screens, trying to fathom how all the numbers fit into Big Bang dynamics. Oh, wait: it's the same now."

"You poor baby," he said. "How was your morning run?"

"No time this morning. I hit the gym for a kickboxing session instead."

M. Carroll, *Plato's Labyrinth*, Science and Fiction,
https://doi.org/10.1007/978-3-030-91709-8_5

Xavier glowered distractedly at the floor between them.

"I know that look, boss. Plain as the face on…your face. You're worried about something." But Xavier was always worried about something, so she added, "Something new."

Together, the two headed down the corridor toward the conference room, where the coffee was. "It's the damned language thing."

"I guess we knew that part was going to be problematic. But you can come close, right? Close enough to understand them?"

"My tutor has his doubts. Even the written stuff is still an open question."

"The Linear A?"

Xavier smiled. "You've been doing your homework. Yeah, they've found the written stuff on pot sherds and such, but if you're a praying woman, pray somebody can break the language code before we have to go."

Kat took a sideways glance at Xavier. He looked careworn, older than his forty-five years. Xavier had trekked more slips than anybody. It took something out of him each time, but he insisted on being the point of the spear.

He said, "For now, we stick to what we know. Modern English sites for you and me, and the French target for Monsieur Todd. He's studying medieval variations now. In his spare time."

"Which I have lots of," a voice boomed through the door, "especially with my imminent Q-slip."

Todd entered the room. His sonorous voice still surprised Katya, so disconnected from his small and wiry stature. It seemed to be a theme with the guys at ChronoCorp: thin and nerdy in an endearing way. A mohawk of mousy brown hair sat uncomfortably on top of his head, as if perched for an escape attempt. A carpet of three-day-old whiskers bristled along his angular jawline.

"Slacker," Katya growled, hooking a thumb toward Xavier. "This guy's working on three languages at once—if you count English."

Todd grabbed a coffee mug. "He definitely needs to work on his English. But for my philosophy major, they made me read most of the works in their original languages: Kant in German, Sartre in French."

"Winnie the Pooh in English?" Kat asked.

"Some of the best philosophy, right there."

"I thought you minored in interior design or something."

"Architecture. I actually had three minors until I finally settled on my major. That makes me either hopelessly ADHD or a well-rounded renaissance kind of guy." He took a sip from his mug and held it aloft, grimacing. "I think this stuff was brewed in the Renaissance. I see you guys still haven't invested in the real thing." Then he peeked into the hall. "Hey Brianne, come on in!"

Kat had forgotten all about the new recruit, who had joined them with squeaky-clean clothes and a squeaky-clean resume, cleared by Xavier himself. Cautious, carefully coiffed, makeup applied to create a subtly elegant effect— in short, Brianne was everything Kat was not. Kat tried to remember if she had even put on any makeup this morning. "Welcome, Brianne," she said in the friendliest tone she could muster.

"Thanks. You're Katya, right? And Xavier." She offered a hand to the boss. "Very pleased to meet you in person."

Xavier shook her hand. "Usually better than looking at someone on a screen. Coffee?"

"Wait a minute," Todd said. "You're not suggesting that this swill is from the brown beans of Columbia or some such?" He peered into his mug as if he'd just found a mummified rat floating in the brew.

"You'd hope we'd be getting used to those pencil shavings by now." Katya glared at Xavier.

The boss shrugged. "You guys want top of the line coffee or bonuses at Christmas?"

Todd sloshed his mug in Xavier's direction. "Don't think we'll forget. Bonuses at the holidays." He started down the hallway with Brianne in tow, whistling *It's Beginning to Look a Lot Like Christmas*.

As soon as the two turned the corner, Katya leaned toward Xavier and lowered her voice. "Can I talk to you privately for a sec?"

He glanced around the room and shrugged. "How's this?" At that moment, two workmen came in with a massive spool of computer cabling.

"How about the Arena?"

So, the two made their way down the corridor and followed another hall to the right, which opened into an anteroom. Using a security key card, they stepped through its double doors.

The first time Katya heard of the Transport Arena, she envisioned a large, dark, oval-shaped chamber encircled by balconies, with spotlights casting beams onto rows of high-tech couches. At times, her mind mixed the preconception with the transporter room in Star Trek, a place of glowing buttons and weird platforms.

But the reality was quite different. The Transport Arena reminded Katya more of her dining nook at home: a cozy, squarish room with one side extending into a half-octagon where her bay window was. In this case, there were no windows at all, but the extended portion of the room rose up three steps onto a raised dais. The little stage held two identical time pods, each as long as a person and reminiscent of cheap metal coffins with 1950s-style toasters at one end. Instead of twinkling lights and strange sci-fi levers, the pods were controlled by two simple laptops attached to larger touchscreens on a counter across the room. On the wall hung four necklace-like objects, cords holding small, metallic devices. The team called them talismans. Once in the field, the devices were the travelers' ticket home. Each was labeled in the literary cadence of Dr. Seuss: Pod 1, Pod 2, Thing 1, and Thing 2. Cables snaked across the floor, held down by plastic sleeves and duct tape for easy access.

All things considered, the room was surreal enough to carry off its Transport Arena moniker. And her vision got one thing right: spotlights bathed the computer work stations and the pods in pools of light, while the rest of the room—clinically white and sterile—was dimly lit. Hanging on the wall beside the door was a small shovel with letters painted on the scoop: *Sometimes it just piles up dark and deep. That's when you start shoveling.*

Katya sat at one of the workstations. Xavier leaned against a pod, arms crossed over his bony chest. "What's up?"

"I think I've figured something out about my trip into the office that day."

"A resounding success! Same with Todd's two forays," he smiled.

Katya traced a figure 8 on the desktop with her finger as she said, "Well, our celebrations may have been premature." She clasped her hands as she leaned into the table edge. "We were shooting for sending me back on a Saturday when no one would see."

"Check. And no one saw, right?"

"True. And that was Saturday the twenty-third. But I was looking at the bulletin board this morning. The one in the conference room." She tapped her finger on the desk nervously. "There's an old announcement of Jeff's party. He didn't put it up until that Sunday morning when he came in for his backpack. Sunday the twenty-forth."

"We really need to clean that thing off," Xavier said, glancing in the direction of the conference room. He seemed ready to spring back down the hall to do some house-cleaning. Katya leaned in front of him, urgency in her voice.

"Xavier, listen. When I went back, allegedly to that Saturday, it was there. I remember it—right there on the board. I asked Jeff, just to make sure. This paper *should not have been here* on the day I arrived. But it was."

Xavier pinched his chin. "We missed."

Katya nodded. "We missed. By a day at least. And I was only going back by five days or so. If the effect intensifies with time, we might miss our target by years."

"Even decades," Xavier muttered. "Decades. That's as old as some of the flyers on our bulletin board. Well how do you like that?"

"Not much."

"Me neither. And we have yet to see if we can send the pods *where* we need them to go. I figure we're two weeks behind schedule as is, and this won't help. We're going to have to do some more number crunching; see what went wrong."

"That should probably be at the top of our list." She felt a tidal wave of guilt pushing her from behind, until finally, she blurted out, "I should mention something else."

He pulled his glasses down his nose and peered over them, meeting Katya's eyes. "Yeah?"

"Remember when Darlene tripped over the cord?"

"She's going to get workman's comp from it. Just found out from HR. Broken wrist on the job."

"That's nice." She waited a beat. "I moved it."

"You *what?*" Xavier stepped down from the platform toward her, his spine straight, eyes wide.

"I moved the cord. But I changed my mind and put it back. Quantum consequences and all that."

"Kat, that was a very dangerous thing to do. We've gone over and over this: we have no idea what kind of long-term effects our actions might trigger."

She wilted. "I know. I'm sorry. But it would have been a minor thing, and it would have saved poor Darlene all those medical bills and that ER trip—"

Xavier was shaking his head already, with gusto. "No, no, no." He pointed toward the staging area, toward the pods, and began to pace. "When we climb into those things, we are Galileo at his telescope, not Mother Theresa at her—her—"

"Hospital?" Katya tried to help.

"Convent. Orphanage. Whatever. We're objective historians, not repairmen. We observe. We document. We are not expending vast resources to fix our own mistakes or right wrongs or ease somebody's hangnail back in the past, when it doesn't matter anymore."

It might matter to them, in the moment, Katya thought, recalling her conversation with her father. *Think where we'd be now,* he had said.

She said nothing.

Xavier leaned forward and planted his palms on the table in front of her. "This is new territory, Kat. We're scholars. We're biographers and chroniclers. Our new funding depends on it. Mila van Dijk has seen to it that we can continue, but we need to play by her rules. Once we make our research public, the one thing we cannot afford is for anyone to accuse us of playing God."

"But that's what time travel is, isn't it? Playing God?"

"We go into a place, watch and learn, and get out before we have any effect. That's our job description. Anything else is tantamount to toying with—" He flailed his hands as if trying to grab the word from the air in front of him. "Toying with eternity. Or something."

But he was wrong. She knew it. She couldn't let it go. "We'll never know the true ramifications if we don't experiment just a little. Push the envelope, explore the borders. We're researchers. And until someone else figures out how to execute a Q-slip, we're the only game in town. It's an opportunity to explore the heart of quantum physics."

"Our financer has given us a different charge. Ancient Thera is our focus for now. Everything has to move us in that direction."

She felt the frustration welling up inside. "But there's a whole lot more involved here than archaeology. We've pulled the curtain back on the entire quantum universe to explore in ways that nobody ever could." Katya realized that her voice had been rising.

Xavier quieted. He was studying her, and she was sure Xavier knew she wouldn't let this go. After a moment, he said, "And I'll bet you have something particular in mind?"

"Todd's supposed to do the test run Friday, but he told me he's nervous about the medieval French he's working on. He feels like he's not going to be ready."

"It will only be a short trip. He probably won't even run into anybody."

"But we've got other issues, too. Other things to work out. Maybe we need to go on another trip that's shorter in duration, closer in physical location, and closer to the present."

"It would be safer to work out the kinks with a shorter Q-slip," he said, warming to the idea. "And I'd be more comfortable with a target less distant on the timeline—a shallower quantum dive—until we understand our variables better."

"Exactly. Why not send me on Friday's test run, and make it to our New York target? Run architecture seven for me, since I've been studying the scenario and all the background stuff. But it won't be glorified sightseeing; this time it will have some teeth. Todd can bone up some more on his old-world French for architecture five—he's been dying to go back to his beloved medieval Europe—and I can do the grunt work for New York. Besides, the techs have the new lozenge imager ready. I have a strong digestive system. I'll take one for the team." She patted her stomach.

Xavier groaned. "You, Kat, *you* should have been in the diplomatic core…"

"And another thing: I've got a top hat."

"A what?"

"A top hat. I'll fit right in at the location of our quantum slip."

"I'm not sure women wore them in nineteenth century New York."

"I read it somewhere. Besides, maybe I'll start a trend. Now *that* could make some quantum waves."

He looked at her with his head tilted. "All right. We'll clear it with Todd, get the camera ready for you to swallow, and shuffle the schedule."

She fought down a grin. "Thanks, Boss. You'll be glad we did. This should get us back on track."

"And no Florence Nightingale stuff, right?"

"Right."

"Promise me."

Katya already had a phone in her hand. "Todd? Xavier and I have a proposal…Conference room, quick."

All the while, she couldn't stop the rambling thoughts about her mother's agonizing passing, and all the agony it caused her and her father for so long after.

What would be the harm?

In fact, who was to say that it hadn't happened already? Leonardo da Vinci with his submarine inventions and diagrams of flying contraptions had a brain well ahead of his time. And what about Isaac Newton? Certainly, he stood "on the shoulders of giants"—but wasn't it possible that he was referring to giants from a future era, and not the past?

If she *could* go back, and if she was not the first, she might not be the last. She had better make it good.

Katya envisioned an hourglass, its sand trickling from the top to the bottom. The top was the future, the bottom the past. But what if she could turn that hourglass on its side? What if the work of ChronoCorp could turn that hourglass upside down for a moment and mix the sand from the past with the sand from the future? She envisioned the glass rocking back and forth, back and forth, as those from the present took knowledge to the past, improving life, speeding progress.

She would start with something innocuous, something that wouldn't make too much of a splash. Cancer research was too risky, too impactful, but why not the obscure work of a British sculptor? Art was subjective. Playground of the rich? And surely paleontology never caused great wars or shook the foundations of the status quo, did it?

Yes: she would begin with paleontology and art, an interesting duo. The New York Q-slip offered both for the price of one, though that wasn't the official plan. But her father had told her many childhood stories of the dinosaur sculptor in New York, and Hawkins would become an integral part of the plan. If all went well in New York, then she would make a more significant move on the next foray. Perhaps she could even convince Xavier to let them start on medical science. Then, it would be on to more scientific advances for the good of everyone.

Besides, if she didn't take the initiative, someone else would, and that someone might have dark intentions. If it was to happen, she should be that someone.

Kat stepped through the double door of the Transport Arena, spotting Brianne a few paces down the hall, walking in her direction. Or was she? The woman's gait seemed to have changed as soon as Katya spotted her, as if she was changing course mid-stride. Had she been listening at keyholes? There was no need; she was an employee with clearance.

"Hey Brianne," Kat tried to sound casual. "Settling down okay? Got any questions?"

"I'm good," she said. But the tech didn't sound good to Katya. She sounded like she'd been caught with her hand in the cookie jar.

6

The Wilds of Wyoming (and NYC)
The Present, Como Bluff, Wyoming

Ajit Joshi scrambled over the crest of a ridge. The wind blew dust into his eyes, but by now he was immune to it. He was embarking upon a different kind of time travel, and he was feeling carefree. He need not concern himself with his forgotten laundry at home; his clothing didn't last in the rugged wilds of Wyoming anyway. It was usually the knees of his jeans that went first.

Across a small valley rose another butte, a twin to the one he was on: pink and white layers of stone atop a mountain of loose gravel and boulders. Scrubby gray-green bushes twisted their way from the ground, but they were of no interest to him. It was what was beneath their roots, pressed between tons of stone and sand. Thar be dragons. This was Como Bluff, and it was full of them. It was a place of prehistory and human history, of Bone Wars, of revelations and missteps. Reed's Quarry 4 was one of the richer beds in the area. Over the century since the region's first fossil discoveries, diggers had unearthed a trove of creatures from another time: top predator Allosaurus *fragilis*, swan-necked Apatosaurus, Barosaurus, Camarasaurus, Stegosaurus. All here, and probably more to come.

At the base of the far bluff flapped several orange tarps, chained to the ground with stakes and rope. A young woman peered from beneath the edge of one and called up.

"Hey, Doc! Got a pes for ya."

He slid down the slope like a surfer, one hand on his white straw hat, and tripped his way across the last few meters of uneven ground.

The grad held out a chunk of rock. A darkened blot edged from its side, the curve of fossilized bone. She pointed to it. "Rounded toes. Not a carnivore.

© The Author(s), under exclusive license to Springer Nature Switzerland AG 2021
M. Carroll, *Plato's Labyrinth*, Science and Fiction,
https://doi.org/10.1007/978-3-030-91709-8_6

But then this…and here." She brushed her finger against the stony form. "We were hoping maybe we got another Stegosaurian?"

Ajit held it up to the light. "Wouldn't that be nice?"

He grabbed a metal tool the size of a pen and scraped a small bit of matrix from the piece. Another dark shape arced along the edge. "Large digit II."

"Wider than III and IV."

"Right. It fits nicely with a Stegosaurus hoof, less uniform than, say, Kentrosaurus. Does this lump come with anything else?" He took his hat off and ducked under the tarp.

Several grad students waved at him. One said, "Doc Ajit, welcome back. Did you bring us peanut butter?"

"Got a whole crate in the truck."

"We've been digging for our supper," said the eager young man.

"Excellent," Ajit said. "Excellent." He said it again, but he didn't feel excellent. Something nagged at the back of his mind. He had finally pushed himself beyond the sharp pinch of his wife's passing, and while it would never go away entirely, at least these days he could file it away in a mental drawer marked "to deal with later." There was exciting work at hand. But still: something else pulled at him.

It was those damned statues. A mere historical distraction. Why did they seem so important? Somehow, he sensed that the sculptures were interlaced with Katya's work. It was a silly thought, but he couldn't shake it. Dinosaur statues in Central Park. What was it Katya had said? Something about the flow of time, of things being linked…

Como Bluff was miles and miles from the nearest cell tower. He had to get to a landline. Medicine Bow was closest. He would take Marshall Road down to State Highway 30, and west until it turned into Main Street. If his cell still didn't connect, he'd hit the nearest phone. There was one in the hotel near the center of town. It was time to get in touch with young Bradley Glenn.

*   *   *

The landline rang out in the darkened room. The detritus of digs, piles of file folders and sketchbooks, drawers full of tiny excavation tools slumbered in the twilight.

It rang again. After a few rings, the voice mail system silenced the phone's little sonata. But Ajit Joshi didn't know what kind of message to record. There was no number to leave, no way Dr. Glenn could get in touch with him. And to what end? A hunch? A conversation to get information without knowing how to apply it? It had been a long trip to the telephone lines in Medicine Bow, and it would be an even longer return to Como.

* * *

Delaney skirted a thorny bush and ducked behind a tree, following Bradley at a safe distance. He didn't suspect she was there. Bradley was headed for the Ramble, or maybe he was interested in Belvedere Castle. His lunchtime voyages were an enigma to everyone, although he sometimes sketched on his break. She was about to solve the mystery.

The New York breeze, seasoned by the flowers of the park and the aroma of a hundred sidewalk deli stands, riffled through her hair. She breathed it in, felt the sun on her face. Bradley turned down a path and disappeared behind a rough hedge. Delaney waited for a few moments, taking a bite of granola bar, giving him lots of room. When it seemed safe, she followed. She would surprise him and then maybe buy him an ice cream cone at Bethesda Terrace.

She rounded the pathway and spotted him ahead. He was approaching a bench. Someone was there, a rumpled man with a tattered hat and a black plastic bag next to his feet. He looked like a homeless guy she had seen in the park before, a man who usually hung out closer to the Carousel. Maybe he was changing his routine.

She stopped in the shadow of a tree and watched. Surprisingly, Bradley sat down beside the man. It wasn't the first time Delaney had noticed Brad's athletic form, his smooth coordination honed from years of cycling. As he sat, the man straightened, squared his shoulders, seemed to gain strength. The two of them looked out across the lake, comfortable in each other's presence. *Curious.*

They spoke for a few minutes, watching the undulating pseudo-medieval reflection of Belvedere Castle in the lake, disturbed now and then by a flotilla of Mallards. The two men laughed and chatted easily. Bradley held out his bag lunch and the man reached inside. Brad took something else out and they dined. Delaney yanked the phone from her pocket to check the time. The lunch break was waning. Perhaps she should head back, but the unfolding scene mesmerized her.

Bradley stood and fished an apple from his bag. He handed it to the man and nodded. The man said something and waved. Her coworker turned toward her. She leaned back behind the trunk of the tree and waited. Bradley walked by. She didn't feel like teasing him anymore. It was Bradley's secret, and somehow it seemed right to keep it that way. He didn't need to know that she knew. Her little spy trip made her feel dirty.

Late in the afternoon, he emerged from behind his monitor and announced that he was headed for the candy machine upstairs. "I'm peckish."

Delaney leapt up. "Me, too." She leaned toward him and lowered her voice. "This one's my treat."

* * *

Satiated with chocolate, peanuts, and gooey caramel, Brad returned to his desk and picked up the phone. His voicemail had a missed call, and this one was no credit card company or announcement of a winning lottery ticket. The caller I.D. had a Wyoming area code. Brad narrowed the number down to the tiny Wyoming village of Medicine Bow. He knew of only one person it could be: his acquaintance from the past, Ajit Joshi. Ajit was a nice guy, a good researcher, a scientist who spent lots of time in fossil sites outside of the little Wyoming town, and a man careful with his time. A phone call from "Bow" must have had an important agenda. A quick internet search marked the call as from the Virginian hotel.

Brad put a call in, but Ajit was not listed as a guest there. He could think of only one other possible connection: Ajit's daughter, good ol' what's-her-name. Finally, he remembered she went by the nickname of Kat, so he'd call her that if he could track her down. He'd lost the number for her last place of employment, a number given him by Ajit in case the paleontologist was in the field and the AMNS needed to get hold of him. She worked at a spot in Fort Collins, Colorado, an outfit called ChronoCorp. They probably made decorative clocks. After navigating several hoops and hitting various numbers, a real receptionist answered and put him through.

He had only met Ajit's daughter once, at an awards banquet for Ajit. Brad was just entering his paleontology masters at Kent State and Kat was headed for her freshman year at Colorado School of Mines. He couldn't remember the sound of her voice or the look of her face. She was an unknown quantity, and she was answering her phone.

"Digital Quantities Department, this is Dr. Joshi."

Damn. She had to answer with her last name. He decided to risk it.

"Hello, Kat? You may not remember me, but this is Bradley Glenn out at AMNS."

After a pause, she said, "Sure, Brad. I remember you. You're the guy who spilled punch all over me at Dad's award bash. But don't worry, I've moved on. Forgotten. Really." There was a good-natured edge to her voice. This was going well.

"I promise not to do that again should the opportunity arise."

"I'm sure we're both wearing more expensive clothing these days."

Oh, the banter was swimming along. "Yes, yes. So I think I got a call from your dad this afternoon, and I was wondering if you knew how I could get ahold of him."

Another long pause told him this might not be so easy. Was she disappointed that he wasn't calling on a social matter? Was she heaving a sigh of relief that he was calling about business? Was she simply searching for his contact information? She should trust him. After all, he was with the American Museum of Nature and Science, a hallowed establishment.

"He's up in Wyoming, working in the north. It's tough when he's in the field, as I'm sure you know. Unless you're a desk jockey."

Desk jockey? That was a bit cheeky. He did work for a museum, but he'd done his share of time in the dirt of real excavations. "I've eaten plenty of sand, believe me. Did some work in Patagonia just a couple months ago."

"Patagonia. Sounds nice."

"It was no vacation. Tents, a lot of cold food and hot days, back-breaking shoveling and the kind of spading and detail work with dental tools that will give you a migraine." He was beginning to whine. Why was he bothering to defend himself? With any luck he would never see this sarcastic woman face to face again. It was time for him to wrap things up. "So anyway, no thoughts on how I might connect?"

"Not a one," she said.

He wasn't sure what to say, but he knew he wasn't going to win any battles taking the low road. He turned his strategy to diplomacy. "Okay, Dr. Joshi, if you could just tell him I tried to return his call—"

"Yes," she said. "Yes, of course I will. Thanks for the call, Bradley."

Apparently, he had been dismissed. He hated it when people called him Bradley. It sounded so much like "badly," which is exactly how this conversation had ended.

7

Blue Glows and Lamplights
Into the Past, Central Park, Manhattan

Katya stood next to the time pod, fuming. She should never have taken the call. Not now. She was sure she had been abrupt with Bradley Glenn, but of all the perfectly lousy times to call! And she knew it was no coincidence that the guy had just contacted her out of the blue, considering her father's recent, and awkward, mention of him. She took a deep breath to calm herself.

All screens in the room glowed, as did everybody in the room--from the anticipation.

"Lasers almost up to power," one of the techs called out. The entire building throbbed in time to the pulsing of the ring lasers, light beams spiraling up the tower behind them. The engineers had disguised the tall structure as a cell phone tower, but it housed something much more complex. Kat was nervous, waiting for her Q-slip—the deepest dive so far—and the geographic target had to be the most accurate to date. With each test, they ventured farther away, both in the three-dimensional world and in the timeline. What if they missed Central Park completely? What if they missed New York City? With an eventual target like the island of Thera, they couldn't afford to miss by much. The traveler would get drenched...or drowned.

She felt her pulse race, her lungs tighten. She was about to sink into that river of Time, into a tributary so few had visited. The current would take her like a leaf, and then she would drift against the flow, upstream and backward, ever backward. Time would run its course in reverse. The sun's looping trail across the sky would move farther south, then work its way northward again with the seasons, summer into spring, blizzarding snows to fiery leaves, eventually to the promise of new life yet again. She could imagine the pages of the calendars and Day-timers and personal journals flipping back upon

© The Author(s), under exclusive license to Springer Nature Switzerland AG 2021
M. Carroll, *Plato's Labyrinth*, Science and Fiction,
https://doi.org/10.1007/978-3-030-91709-8_7

themselves, Januaries to Decembers, long entries shrinking to blank pages. Scattered stones stacking themselves into palaces and castles, and the shriveled mummies dropping their wraps, their skins blossoming from desiccated browns to the flush of life.

To calm her nerves, she went over it all in her mind again.

Gravity affects light.

Gravity affects time.

Light also affects gravity and, consequently, time.

The powerful light of the ring lasers built to a crescendo in a tightening helix through the tower. The beams were silent, of course, but their effect—the twisting of the very fabric of space/time around them—radiated a low rumble that she could feel in her chest. In physics, they had a name for this twisting of space and time; they called it frame dragging. But in practice, in the experience of this mighty experiment, she called it fearsome. The quavering reverberation brimmed with the energy of the Creation, the exhilaration and wonder of those first moments of existence as the universe—everything that had ever been in it, everything that ever would be—flashed into being.

The new hire, Brianne, looked around and then turned to Xavier. "She's not bringing any electronics along?"

In the tension of waiting for the system to power up to capacity, Kat sensed a teachable moment, and she took it. The waiting was killing her, and it would be good distraction. "There's a hitch to that, and it gets kind of embarrassing."

"Oh?"

"Here's the really weird thing," Todd injected himself into the conversation. "You can't take anything with you on a Q-slip, and you can't bring anything back. You take some kind of recording device, and it won't work. In a few hours it will completely vanish. Same for electronics, or really anything artificial. Only living tissue, stuff that can repair itself on the fly, remains intact."

Brianne's perfect mascara exaggerated her widened eyes. "What about clothes?"

Kat cleared her throat. "That's a real problem. They're okay for a while, long enough—hopefully—to find something local that will do the job."

"What if you rushed off somewhere, some *when*, did a quick recording, took a few snapshots, and came back before everything fell apart?"

"More quantum weirdness," Xavier said. "You bring back jpegs, electronically saved to a stick or flash drive, and print them out and they disappear in hours, along with the data on the stick. They erase their own little pixels from hard drives. We even tried taking a photo of the original jpeg printout, and they both vanished. It's crazy. Some entanglement thing. Anything having to do with the era we've traveled to is linked. It all goes away, except for memories. In fact, sometimes cameras or recording gadgets simply won't work in the field. They get scrambled on the way back."

"The only thing that doesn't change is the pod we travel in," Kat said, "and the talisman we carry as the triggering device." She held up the small fob on a lanyard, her high-tech necklace, and let it drop back beneath her blouse. "But it duplicates, so that the pod here remains here, and the traveler disappears into the other time, but we materialize in an identical pod there."

"Two pods. Weird."

"We're assuming that upon the traveler's return the second pod disintegrates into the quantum field. But as long as there is a person from our time in another era, a pod will materialize to bring them home."

Todd raised an eyebrow. "We hope."

"Yes, but it makes sense," Kat said. "The problem is one of conservation of mass. As long as one of us is in another time, the universe is unbalanced. There is more mass in the era we are visiting than there should be. The tendency of nature is to balance things out, so the natural universe wants the traveler to return to the world, the time, they came from. All we have to do is trigger the talisman and wait for a pod to materialize."

"Wait," Brianne objected. "So the pod doesn't go with you?"

"From your viewpoint, I'm going to climb into that pod and disappear. But to me, the traveler, I'm inside the same pod as the surroundings outside of it change, and then I climb out. The pod I came in—what we call the ghost pod—eventually vanishes into the quantum field, until we call it again with the talisman."

"That's why we call it a ghost pod," Todd added. "It's not really there. Technically."

Xavier spoke as if calling a meeting to order, his voice cracking slightly. "Don't get too caught up on the details. This will be a short trip, won't it, Kat?" He said the last with emphasis.

She held up her hands in surrender. "Don't worry, don't worry. Short and sweet. No more than ten minutes, as agreed."

Kat climbed into the pod. As Todd sealed the lid, Katya glimpsed the newbie, Brianne, looking quite concerned.

* * *

Umpire Rock may not have been the most comfortable of places to sit in Central Park, but Hawkins loved it. The boulder was like a great throne, scored by layers of time. *Time.* There was such a heaviness to it—a formidable, relentless force ruling over the universe. Time was a watercourse, conveying people and things and places. Sometimes it rushed at breakneck speed. At other times, like tonight, it drifted lazily. He thought about the dinosaurs drowsing under his feet. Not the ones he had made and lost, but the ones the Creator had fashioned so long ago. How would they look as they lumbered across the Sheep Meadow, down past the carousel and on to Columbus Circle? Hadrosaurs in the Heckscher grounds! Teleosaurs in Shakespeare Garden!

He found comfort sitting on this natural throne of the ages. The light was failing now, and the lamplighters had not yet come. Looking to the south, toward the pond, he could see no hint of the disaster, no clue that his fallen fabrications slumbered beneath the sod. Iguanodon, Pterodactyl, Megalosaurus, all extinct for a second tragic time.

Time to go home. He was being prudent: what if the thugs knew he was back in town? What if they were waiting to do more than bludgeon his concrete creations? What if they were out for him? Hawkins wouldn't put it past the Tammany Hall gang.

This neighborhood at this time of night was enough to give even able-bodied rugby players the shivers. He could cut across the park; the path to the north would spit him out onto 65th, just a couple blocks from his hotel.

He turned back into the commons, nearly losing his footing against a raised flowerbed. There was no moon. That might be good protection against werewolves, but it was lousy for an evening hike through Central Park. Up ahead, beyond the Heckscher Ballfields, the path turned northwest along the Sheep Meadow, into a dense stand of trees. Gas lamps lit some of the way, but eventually the path darkened until he could barely see the trail. Just as he was losing his nerve, he spotted a faint blue glow through the trees. The light held an odd quality, unlike the lamps he had passed moments ago. This light seemed less sure of itself. Less cemented in reality. Ethereal.

He squinted toward the source. Amorphous shapes flickered and undulated in the radiance, but nothing recognizable. The scene was unusual. Hawkins thought about what his dog would do in the situation: her hackles would be raised, certainly.

He heard a voice issuing from within, but he couldn't catch the words. It was a high voice, like that of a young boy. There was no content, just tone.

"Mr. Hawkins?" The voice came again, and this time the tinny echoes focused to clarity. Not the voice of a boy, Hawkins reconsidered, but of a woman. "I want to show you something. Something about your fine work. It's something that no one knows about at this particular time."

When the woman of the shadows spoke of "this particular time," Hawkins got the distinct impression that she was not speaking euphemistically. No, there was nothing allegorical about the term, but rather something literal.

"And why would you do such a thing?" Hawkins asked into the dimness, trying to keep his tenor steady. Her disembodied voice disturbed him. He called out, "Show yourself."

"Call it a favor for an old friend." Her voice was clear now.

"If you'll pardon me for saying so," Hawkins offered tentatively, "you don't sound old."

"Well, now, that kind of thing is relative, don't you think?" There was something in her voice that hadn't been there before: amusement.

She stepped from the shadows, the blue light fading away around her. At first, Hawkins thought her exceedingly tall, but as her figure emerged from the glowing fog, it became apparent that she wore a top hat, very strange fashion for a modern woman. She carried some sort of valise.

"My name is Katya Joshi. And you are Benjamin Waterhouse Hawkins?"

"I certainly am." He kept his tone cautious.

"I understand you've been creating some sculptures for Central Park," Katya said.

The events surrounding the Paleozoic Museum were still fresh, even after all the time. "That was years ago. I've been to England and back again. I'm at the College of New Jersey now. They treat me far better than our dear Tammany Hall leadership did. The New York political machinery has no influence over an institute of higher learning in New Jersey."

The woman furrowed her brow and studied the ground. "College of New Jersey. Princeton. Five years wrong?" She looked up and met his eyes. "May I ask you a strange question?"

"In light of your entrance, nothing will seem strange, madam."

"What time is it?"

"Twelve minutes past seven of an evening." Not such a strange question.

"What year is it?"

A chill ruffled the back of his neck. The scientist in him raised a red flag of skepticism, but the observer in him urged him on. "It is the year of our Lord eighteen hundred and seventy-seven."

"And you left New York in?"

"After we abandoned the Paleozoic Museum of Central Park, I lectured." His mind returned to the time with fondness. Strange how in the midst of tragedy and injustice, one could have fond memories mixed in. "Yes, lecturing. And I did some dinosaur skeleton reconstruction for the fine Smithsonian Institution, but I returned home in eighteen hundred and seventy-four."

"And then back here again."

He nodded. "To America, yes. Almost immediately. Not to here, but to New Jersey. I'm just visiting. Is it important? The date, I mean."

The woman bobbed her chin slightly. "It could be. I don't know." She brushed some lint from her sleeve, and the cuff tore away. Hawkins marveled that this woman would bother to wear a fine top hat and yet not invest in a respectable coat.

The lamplighters were making their way down the path now. She motioned him to step into the light of a nearby lamp post. She opened her valise and pulled out some papers. The sheets seemed tattered, like her entire ensemble.

"Can you tell me what you think of these? They're reconstructions of a Hadrosaur, and an Iguanodon, and a Tyrannosaur."

"Tyrannosaur? I am not familiar with that one."

She handed the pages to him. The paper was slick, like nothing he had seen. It crackled as he leafed through the drawings. And what drawings they were! Quite different from his reconstructions. His giants were based on slow-moving lizards and other primordial reptiles. But these? These held their heads low, level with the backbone. Their tails stretched out behind them, with their hind legs bent, as if they were greyhounds ready for the track. And the climate around them was alive with cycads and ginkgoes, with arroyos and buttes. The skies flashed with lightning and glowed in sunrises.

"Where did these come from?"

"Do you like them?"

He nodded pensively, not taking his eyes off the Hadrosaur. "Very much indeed."

She pulled a small case from her pocket. It was no larger than a tobacco tin. From it, she drew a tiny capsule. She waved it around strangely and then did something remarkable. She popped it into her mouth and swallowed it. She returned the case to her pocket and said, "I wish I could leave them with you, but I've got to get back."

"Who did these? And what medium?"

"Several artists. They had—" She paused, obviously debating just how to put something. "They had information unavailable to you."

A sense of urgency rose in his throat. She seemed restless; she was at the point of departure. This singular opportunity was slipping away. "Where can I get this information? See the fossils that these craftsmen used?"

"You have the ideas—you can keep those." She tapped her temple beneath her top hat. "Learn from them."

The valise cracked as she slid the drawings back in. Hawkins followed them with his eyes as long as he could. As soon as the drawings disappeared into the old case, Hawkins looked back at his odd companion. She was fingering an unusual pendant, a metallic chevron on a lanyard. His gaze fell to her blouse. It was in tatters, and he could see little patches of her skin through it. He averted his vision as any gentleman would.

She reached her hand toward him boldly. He knew Americans were self-assured, sometimes to the point of rudeness, but this woman did not strike him as insolent. She certainly came from a place with strange customs. He shook her hand.

"Mr. Hawkins, it's been a great pleasure."

"The pleasure is mine, Miss Joshi."

She looked around at the dark settings of Central Park as if it was a new world to her. Turning abruptly, she strutted a few paces down the walk and into the trees. He frowned. Was there any truth to what she had said about the ancient creatures?

What was he thinking? He called after her. "Miss Joshi, please allow me to escort you. It is not safe here at this hour."

She didn't reply. That ethereal blue glow flashed again, lancing out through the trees, casting shafts of light against the pavement around him. It blinded him for a moment. He looked away in time to see his own shadow, alone against the hedge on the opposite side of the walk. And he knew she was gone. He needed to get back to New Jersey. He had some sketching to do.

8

Dreams of Greatness

Katya sat down next to Todd Tanaka in the murky room. His face was gaunt, his clothes in tatters. He was filthy.

Gazing at the floor, he mumbled to her, "I'm just so tired. Tired of scrounging around for stuff to eat and hiding in buildings instead of enjoying the sunshine."

Somebody had a room monitor set to a news channel. The commentator was saying something about genetic engineering.

"Of course, no one was naive enough to break the rules about dinosaur DNA. Only samples of small species have come back, but as we all know too well, that's been quite enough for disaster. Governor Delacroix has said that the city is living in a state of siege. The swarms of dinosaurs have eaten all the rats in the sewers, just as they did our pets. Those chicken-sized Compsognathus hordes have come out from beneath the streets and turned on the humans. The New York City Sheriff's Department announced this morning that the remains of another homeless person have been found in Central Park."

Todd reached for a remote and shut the thing off. "Guess we asked for it. Bringing back DNA; what were we thinking?" He turned, dissolving into the darkness like a ghost.

The room went cold. Katya was in her bed again. Her pillow slipped to the floor, and the entire room shifted. She shivered and pulled the blanket firmly around her shoulders, then sat up, her nightshirt soaked in sweat. Again.

She hoped the surreal dreams weren't a feature of the Quantum slips, but so far, they had haunted her at each return. Her visit to nineteenth-century New York had gifted her with headaches and night terrors for a week. Todd had mentioned having bad dreams since he went back. Kat wondered about Xavier.

She swung her feet to the floor and padded into the kitchen. Grabbing a mug, she sloshed some day-old coffee from the pot, dumped in some cream and honey, and plopped it into the microwave. What would Waterhouse Hawkins have thought of a microwave oven?

He'd been remarkably receptive, and she had carried out her little experiment well. But how would she know if anything had altered? Some theories said that if something in the past changed, the viewer simply entered into an alternate universe where those changes had always been the norm. Or maybe the agent of change was the only one who would remember the universe as it was before. If that were the case, perhaps paleontology was forever changed, but only she would know. And then what? Would it be safe to go back armed with medical knowledge? To travel to the Gross Clinic of the 1800s and tell them about immunotherapy? Give them formulas for chemotherapeutic drugs? Tell them to wash their hands before surgery? Would she return to a world—a *universe*—where her mother still lived?

The microwave beeped. She grabbed her cup, sat down at the table and rubbed her temples. A bird screeched outside the window. It sounded a lot like Katya imagined a Compsognathus might sound.

The team would need to be responsible, cautious, and pragmatic with their new and terrible power. They must be careful.

She must be careful.

9

Princeton Papers
The Present, Princeton, New Jersey

Spring cleaning at any university was an unwelcome affair for the students who were conscripted to take part. A university as old as Princeton had the kind of dust that made the workers lose their appetites. Chelsea Mayele was a first year post-doc in the Geosciences department. Her focus was the Rhaetian Age of the late Triassic, but for this job, it didn't really matter—she had a pulse and was available for serfdom.

While some of her colleagues had been tasked with moving crates and sweeping decades of dust from concrete floors, her task was to sort through the old files, papers, and blueprints that nobody wanted any more. They had given her a very large dumpster, an electric drill with screw drivers, and a crowbar, and set her to work in the murky basement of the Geosciences Department's ancient main building.

Even Chelsea's keen young eyes couldn't cut through the darkness of the cavernous lower level. She had found a light switch on one end of the vast room, but the lights it triggered were far from her current perch. She had to resort to a flashlight for sorting. The limited beam revealed a mountain of crates and boxes stacked in the center of the long room, its form reminding her of a primordial creature's backbone.

Chelsea had just closed the lid on the third box of old invoices, receipts, and order forms for things like desk pads and inkwells. She wiped her forehead, bemoaning the repetitive contents of everything she had sorted.

"This is grad student abuse," she mumbled to herself. "Slave labor. That's what we are."

She turned to another crate, this one long and shallow. A new shape might be more interesting, at least. Across the top, someone had stenciled the words

© The Author(s), under exclusive license to Springer Nature Switzerland AG 2021
M. Carroll, *Plato's Labyrinth*, Science and Fiction,
https://doi.org/10.1007/978-3-030-91709-8_9

COLLEGE OF NEW JERSEY. It had been ages since Princeton went by the name. More than a century? This was a very, very old box. The thought sent shivers across her shoulders.

There were screws on the top of the container. She attacked them with her drill, but most of them stripped out. Not to worry: all of these boxes were basically glorified trash bins. She jammed the crowbar under the lid and jerked. Splinters flew and the top clattered to the floor. Whatever was inside had been covered with cotton batting, now moldy with age. She put on a new set of gloves, ready to dive into yet another sequence of stratigraphy. Beneath the cotton lay layers of wax paper, and under them...what? The subdued warehouse light showed a warped piece of cardboard and a large paper taped to it along the edges. She aimed her flashlight onto its face. The tape was nearly gone now, merely a yellow smudge in places, but the paper still adhered to the mildewed cardboard. The center of the page was what got her attention: a beautiful Victorian landscape with a giant dragonfly, a clumsy lizard-like creature crawling onto a lakeshore, and a bat-like silhouette against a dusky sunset. Despite its age and the dim light, the scene glowed. In the corner she could make out three letters, BWH. The initials meant nothing to her.

She pulled out the large sheet and laid it carefully on a table. Returning to the crate, she blew dust from the next layer. Beneath it, slumbering in this box for decades, lay a folio of drawings. Many were rough, but some had the polish of the professional artist. One leaf had two names at the top: W. B. Scott and B.W. Hawkins. William Berryman Scott's name was carved onto the wall outside of this very building. But Hawkins? She remembered that name. It was buried in her brain beneath all the data crammed in there for tests and term papers. Hawkins. Victorian artist who sculpted dinosaurs before the name dinosaur had been invented. Could these be originals? She pulled her phone out and did a quick search. Sure enough, Hawkins had spent time at Princeton, back in the nineteenth century.

Suddenly, the dumpster was the last thing on her mind. These loose leafs might be treasures. Should she call someone? What if there were only receipts and work orders further down? She had to dig some more. Ironic how this morning she had thought digging in a Montana butte would be preferable to working in this basement. Now, she was on a different kind of dinosaur dig, and it was plenty exciting.

She missed lunch. She skipped her afternoon lab. It must have been late when Professor Markham peeked in on her progress.

"Miss Mayele, do you know what time it is?"

"Haven't the slightest. Wait until you see what my excavation has uncovered." She held up the color piece, and then a sketch of a theropod. "Waterhouse Hawkins. Originals, I think. Are they important?"

Yes, they were important. She could tell by her professor's face. Markham poured over the drawings using his phone's flashlight. He sorted, compared, scrawled notes on a scrap of paper, and bobbed his head around a lot. All the dinosaurs had a heaviness to them, a primordial inertia to be overcome. These were cold-blooded, overgrown reptiles dragging their tails, hissing toothy snarls, glaring with their malevolent eyes and scaly skin. All but one. Chelsea held it up.

Markham's eyes glowed as he held it in his hands. "I don't understand." He gaped at the corner. "B.W.H. But look at it!"

The sketch was of a hadrosaur, a lovely duck-billed beast. But this beast had a different feel to it. It leaned forward, nearly on all fours, like the other drawings, but this one shifted its weight toward the back legs, its right front paw raised slightly, holding his tail erect like that of a kangaroo. It held its head upright but low, in line with the backbone, ready to spring from the yellowed paper. This was no giant lizard. This was a true dinosaur. Across the drawing, in wide red pencil, someone had scrawled the words, "Not reptilian enough."

"Do you think it's genuine?" Chelsea asked.

"If it is, it's significant. Look at the posture…the gait…even the textures…"

"Looks pretty modern to me," Chelsea said doubtfully.

"Yes," Markham whispered, tapping the edge of the board. "This—this is good."

"Should we call someone? Someone official or something?"

He looked up at her. "Yes, there's a fellow who works at AMNS. Bradley Glenn. Waterhouse Hawkins is a sort of specialty of his. Oddly enough, they recently found some pieces of Hawkins' dinosaur statuary in Central Park, and he was there. I'll have to look him up."

Chelsea stretched and rubbed the small of her back. "Gee, people are digging up all kinds of things lately."

"You might say," Markham muttered, distracted. He sat up straight and turned to Chelsea. "Very good job, Miss Mayele. You've certainly earned the night off. If you want to assist me in cataloging these tomorrow after class, it would be quite helpful."

"Of course," she said, covering a yawn with the back of her hand. "Good night."

"Oh, and Miss Mayele, let's please keep this confidential until we can verify it."

"Yes, of course."

"That means no roommates. No boyfriends. No mom and dad. Just for a little while, good?"

She nodded, stifling another yawn, and headed to get some fresh air.

* * *

A little while turned out to be less than a day. The secretary in the Dean's office overheard the Dean talking to Professor Markham the next morning. By the time the Dean made it over to the Geosciences building to see what the fuss was all about, reporters from the student paper were already there. When Chelsea made it in an hour later, the building was crawling with students and passersby. Markham stepped over to her, brushing aside a cub reporter and an aspiring photographer. The story would be online in minutes.

"This is gonna go viral," Markham grumbled. "We should never have brought these things out of the dungeon, out where people could get to them."

The Dean sauntered up, looking satisfied. "You know what they say: any publicity is good publicity. This just might net us a grant or two."

After taking a cursory glance at the drawings, the Dean left again. Markham spoke to Chelsea confidentially. "These administrators. Their tenure kills their joy of learning. More often than not, it's the fundraisers and politicians who become the deans. They forget—" Markham stopped himself, lunging toward several drawings spread on the table. "Knock it off, you guys! This stuff's fragile. I'll be happy to move things around for you." He held up his gloved hands.

The little gaggle of school journalists left after the excitement had died down, only to be replaced by a larger gaggle of faculty and a few philosophy students from next door.

"Miss Mayele," Markham said, "can you keep an eye on things? Keep everybody from putting their grubby paws on these pieces of history?"

"Sure thing. Are we going to catalog them?"

"I think today we'll be satisfied with simple preservation. We need to get these treasures into safekeeping. At least into a more controlled environment. Call Evie Long."

"The archivist?"

"She'll know what to do."

"I'm on it," Chelsea said, pulling on a pair of plastic gloves.

"I have a phone call to make."

Markham headed back toward his office. He hoped he still had the personal number for Bradley Glenn. The man was needed.

Here.

Now.

10

The Chase

Thomas James, Private Investigator, turned down the alley across the street from the warehouse. His employers worked out of the crumbling hulk of a building—not so crumbling on the inside—and he wanted to put as much real estate between himself and them as he could.

He'd been a good contractor, and had given them what he considered solid intel on ChronoCorp. He could have dug further, but instead, he dug a bit into the darker background of his own employers at the Primus Imperium. He didn't like what he saw, so he set up a meeting. That was his mistake.

They assigned the somber man to give Thomas a sort of tour. Accompanying the man, a cowboy-hat-wearing thug who spoke little, were two strongly built guards. The tour they gave him confirmed all his fears. He had carried out research for clients doing shady business in the past. That was certainly not beneath him; it came with the job. But this was crazy. Insane street on the main street. They were trying to impress him, but all the excursion had done was to convince him that he had to get out. But he had seen too much, and now he feared they were after him. He didn't see anyone just yet, but he felt like a hunted gazelle. Or maybe a mouse beneath a diving hawk.

The scene of those so-called "temporal capsules" gave him the creeps, and when he finally put together what they were for, what the Imperium's ultimate goal was, he ran like a rabbit. He had to get out. He had to let someone in charge of something know what was going on inside that big building, which wasn't nearly as abandoned as it looked. But who could he go to? The local police? The FBI? He actually had a contact at the NSA. Randy was only a pencil-pusher, but he must have had contacts there who would know what to do.

© The Author(s), under exclusive license to Springer Nature Switzerland AG 2021
M. Carroll, *Plato's Labyrinth*, Science and Fiction,
https://doi.org/10.1007/978-3-030-91709-8_10

He had followed instructions, parking in an alley two blocks away. A frigid rain began to fall, making the faintly lit alley even darker. The brickwork walls of the buildings on either side of the pavement reflected on the moistened asphalt, like another universe mirrored across the cracked street. Ahead, the alley took a turn to the right, reinforcing the feeling of a maze—with James as the lab rat. He had the cell phone that the Imperium had given him, but it didn't have his personal contacts in it. For those, he had to get back to his car and his other phone.

He rounded the corner. His car was at the far end, but the way was blocked by a gray SUV. He skidded to a stop, splattering water up his pants legs and almost losing a shoe. A man stepped from the SUV's driver's side and called out, "Mr. James, the Ambassador wants to have a chat with you. He just wants to talk."

James was certain that someone would be talking, but they might be doing it with the butt of a pistol. He preferred not to have that kind of conversation. He spun around and skirted the corner again, heading in the direction he had come. He was beginning to think he wouldn't get out of this in one piece. Now it seemed more important to get the news out. The Imperium was playing with technological fire, and what they intended was not for the good of the world.

At the far end of the alley, between him and the warehouse across the distant street, another SUV pulled up to block the way. Two figures stepped out of it and began to come his way. He looked up. No fire escapes, no windows to crawl through. A line of vents stuck out of the brickwork, but they were a good two stories up, beyond even the best NBA player's jump. Fighting the panic welling up in his chest, he scanned the area for a way out. Between him and the two approaching gorillas, a lone manhole was inset into the pavement. Could he make it? Maybe. He ran.

The assailants were too close. He paused before he got to the manhole, locking eyes with them.

"Where do you think you're going?" one of them said. The thug had no gun or weapon. His tone was menacing enough.

James began to back away from them. He heard a subtle scuff on the pavement behind him, felt a sharp pain at the base of his neck. The alley filled with painful light. Then, oblivion.

11

Missing the Mark

Xavier Stengel squeezed the bridge of his nose between thumb and forefinger, as if trying to keep his forehead from sliding into his chin. "Disturbing," he mumbled. "Damned disturbing."

What's disturbing, Katya thought, *was all three of them huddled around the small desk in Emil's lab.*

Todd pulled up a page of equations on his tablet. "So we're saying that our aim point was seven *years* off."

Xavier tapped his chin with his index finger, a habit when he was annoyed. "The models must be wrong somewhere."

"That's pretty irritating," Todd murmured.

"Or terrifying," Katya said. "I mean, we go back to someplace like Thera a few years late and instead of observing Minoans, we end up buried under yards of lava."

"Well, pumice," Todd said, his eyes locked to his screen. "Packed volcanic dust, pretty much like fluffy concrete."

"Searing fluffy concrete. Not a good place for asthma," Xavier suggested.

"Or breathing of any kind," Katya added. "And if the equation is on a curve, as Emil suspects, then the effect might be magnified the farther back we go. My Q-slip to the office five days' past was off by hours. The trip to the early 1900s was off by seven years. By the time we hit ancient Greece, we could miss our mark by centuries."

"At least we got the location right," Xavier grumbled.

Todd's face darkened. "That's one thing I don't get. We wanted Kat to meet up with Hawkins as part of the test. Cause and effect and all that. But we missed."

© The Author(s), under exclusive license to Springer Nature Switzerland AG 2021
M. Carroll, *Plato's Labyrinth*, Science and Fiction,
https://doi.org/10.1007/978-3-030-91709-8_11

"By a lot," Katya said.

"So why was he there at all? In seven years, he could have been anywhere else. And as he said, he had been to England and back again, and Kat still finds him?"

Xavier cleared his throat. "I think this has to do with what I suggested earlier. If the flow of time does move in currents and eddies, those flows may keep bringing things upstream in a pattern. Our aim point was a spot in the timeline where we would find Hawkins, and we did find him, but not at the point on the timeline we intended."

"So we hit the target in three dimensions, including the person we were after, but missed in the fourth dimension," Katya said.

"It's as if the natural unfolding of the universe keeps that parallel going. We had our coordinates set to meet Hawkins. We missed the time. But in a sense, we got the location right. Central Park, with Waterhouse Hawkins."

Emil stepped through the door. "I hate to be the bearer of bad news, but our new system didn't work."

Katya moaned. "No images? None at all?"

He shook his head. "Sorry."

Todd held up another imaging capsule, grinning at Katya. "At least I don't have to retrieve one of these."

"Fat lot of good it did us," Kat laughed, patting her stomach. "I don't envy anybody trying to find something like that gadget in—in—"

"A stool sample," Todd offered. "So it's still the same problem. Something about transporting technology. Shielding it in organic, living tissue has no effect. We need a new strategy."

"We do have some ideas," Emil said.

"Ideas are good," Todd said. "Good ideas are even better."

"We'll need some…" Emil looked momentarily at a loss for words, and then said, "meat."

"Meat?" Todd asked. "Like hamburger?"

"Like reprogrammed DNA. Building something from scratch. It's like Xavier said. 'Only stuff that can repair itself on the fly remains intact.' Wasn't that it?"

"Yes, Dr. Frankenstein," Xavier said. "Work those ideas. Will you need to wait for a lightning storm to power your laboratory?"

"What we'll need is a team who can splice DNA. I have contacts in industry who might be available." The tech left.

When he was gone, Xavier crossed the room and gazed out the window. "Something's not adding up. This is very unsettling." He seemed to reconsider. "It all comes down to numbers. Math. We can solve this."

"No, Boss," Todd objected. "It's not just a bunch of numbers on a screen. Those figures mean something, and at the end of the day, what they mean is that we've got a whole lot of variables we don't understand. It's dangerous. Maybe dangerous on a reality/universe-wide/what-have-you scale. It seems to me we have a moral obligation to be responsible with this power we've got running up and down that tower out back."

Xavier smiled at Todd, not unkindly. "ChronoCorp's own friar. Everybody needs a moral compass. I'm glad you're here."

"I'm not sure I signed on to be anybody's Jiminy Cricket," Todd said quietly.

"No, you're right. We need to be careful. Ripples in the pond, Todd."

Katya's heart skipped a beat as she remembered her exchange with Waterhouse Hawkins.

"When are we pulling Bradley Glenn into this?" Todd asked. "He's the one who can tell us if anything has changed."

Bradley Glenn, wonder boy. Katya had to admit that she was predisposed to not liking the man. Her father's zeal in getting them to meet made her feel like the adolescent schoolgirl heading for a blind date. And what could Glenn tell them? Had she put into motion something terrible? She felt a trickle of sweat meander down her back. Her face was hot. She needed some air. She was beginning to feel her time-tweaking plan was the worst idea she'd ever had. She made for the door.

Xavier kneaded his hands. "Whatever we do, don't mention this to van Dijk."

"Van Dijk?" Katya stopped short, her breath caught in her chest. "As in Mila van Dijk?"

"She didn't hear," Todd told Xavier. "Must have been when she was at lunch with her dad."

She shoved her fists against her hips. "Oh, fine. First time I take lunch in a month and I miss something big?"

Xavier said, "Not that earth-shaking, Kat. We're invited to our benefactor's private party. And it's not the kind of party one turns down."

Todd's eyes sparkled. "It's at *The Mansion*. Van Dijk's estate. Wednesday night. For grilled salmon, catered by Le Fauve."

"Wednesday? I've got—"

Xavier cut her off. "Cancel it. Whatever it is." It wasn't like him to be so adamant. This must be important.

"Is she going to be grilling us along with the fish?" Katya asked.

Xavier scrunched up his nose. "I don't think so. I think she knows we're getting close to an operational foray, and she wants to share her vision with us."

"Wants us to catch the Minoan fever, as it were," Todd added. "I hear the food is great, but the guests have to pay. Not with money, of course."

On her salary, that got Kat's attention. "Pay?"

Todd closed out his tablet. "We pay for a good dinner by being lectured to."

"The woman is paying your salary." Irritation tinged Xavier's tone. "And mine. We can listen to our patron for an evening."

Todd shot to his feet. "Hey, I know. We could step into a pod and send ourselves to the next morning. We'd be full of good food, fine wine, and already past van Dijk's monologue. A practical use of our technology!"

But Katya was staring down the hall, toward the arena. "And miss the process of mastication, of tasting all that gourmet food? The heartburn? I'm kind of curious to hear what she has to say, maybe witness her wheels turning."

Xavier slapped his hand on the counter. "That's the spirit. Todd can get liquored up on Mila's communion wine while you and I enjoy an evening of ancient Greek history."

Todd held up is hand. "Well, now that you put it that way, maybe I'll just have a glass, and then some coffee. History is always good." He looked toward Xavier, who was smiling.

"Knew you'd come around. Be here at seven o'clock sharp Wednesday evening. Todd, no tee shirts. Kat, no jeans. This is a fancy affair. They're sending a limo. And I think it's best we keep the specifics of our progress close to the vest for now. She will be less inclined to want to supervise."

Todd frowned. "But isn't she technically the supervisor of our missions by definition? All her money must earn her that much, at least."

"Remember that we're the good guys. We're in this for exploration and history and the betterment of mankind."

"Humankind," Todd put in.

"Well put," Katya said.

"Not to mention the preservation of the American way," Todd added.

Katya nodded. "Salvation of the known galaxy." She began humming *America the Beautiful*. Todd joined in.

Xavier waved them off. "But there are other companies out there, other parties, with different agendas. We're not sure what those cover, but you can bet some of them are less than noble."

"Filthy lucre comes to mind," Todd said.

"I could use some filthy lucre," Katya told Todd. "Or even reasonably clean cash. Can I borrow a twenty?"

Xavier took on a more urgent tone. "Look, team. What I'm saying is that secrecy is of the utmost importance. This is proprietary work."

Todd shrugged. "That's what it says on our contracts."

"Sure, but we've got to live it out. It is imperative, *absolutely imperative*, that our work here be kept under wraps until we've had success and detailed reports from Thera. Until then, we keep our heads down and stay below everybody's radar."

"No little stories between even friends or family."

Katya felt a pang of guilt at her recent lunch—and slipped revelations—with her dad. "No stories," she nodded.

"No stories," Todd confirmed.

"Everything, but everything, must be kept within these walls. Our secrecy is…"

Todd and Katya answered in unison, "Imperative."

* * *

Brianne slithered away from the doorway and tiptoed down the corridor. She had a knack for knowing when to quit. Wouldn't want her new employers to get the wrong idea, now would she? In fact, it was time to head home for the day. Maybe make a few calls, send a couple texts, tie up some loose ends and then relax.

12

On the Trail

Rex Berringer knew something was up at ChronoCorp. His client had told him so. He sat behind a slightly misassembled IKEA desk; the furniture bristled with screws and pegs jutting out at odd angles. He had put it together himself. Behind him hung awards from dubious organizations, and a golden nameplate stood on the desktop at center stage, just in front of his Samsung knock-off computer screen.

He pulled up letterhead on his monitor. Across the top of the digital page read the stately words BERRINGER & CO, PRIVATE INVESTIGATIONS. The "and company" part was now his cat, Fifi, as he had fired the secretary last week for insubordination. He liked the thrill of being in business for himself, the flexible hours, the handing out of his business cards at the local BBB meetings. He just didn't like the paperwork. Or the quarterly taxes, which he sometimes paid. He would have to get another secretary soon.

Rex Berringer liked his name a lot more than the one his parents had given him. Melvin Dolp would not have had the same effect on those business cards of his. But Rex? There was an exciting name, the name of kings and of big dinosaurs. It was the moniker of a man of action. He had grown the only successful facial hair he could for the job: a wispy moustache that he dyed to make it more visible. And *Berringer* reminded him of a compact, powerful firearm. It sounded better than Rex Crossbow or Rex Grenade or Rex Bazooka. It was almost perfect.

It would do until he could think of something else.

But for now, Rex Berringer was on the trail of something fishy, at least according to his customer, a Mister Dresden. (Rex Dresden? Maybe not.) Mister Dresden had not divulged what his role at the company was, but those

© The Author(s), under exclusive license to Springer Nature Switzerland AG 2021
M. Carroll, *Plato's Labyrinth*, Science and Fiction,
https://doi.org/10.1007/978-3-030-91709-8_12

he was working for did something related to research or science or some high-tech thing. Rex had tracked down their website, and although it was a bit evasive on the details, he had concluded at least that much from the splash page: gears and equations and graphs behind bubbling beakers. Oddly, there was a subtle flag behind it all, but he couldn't quite make out the details. Mr. Dresden had advised him to not ask too many questions. Apparently there had been a PI before him who had, and the guy had lost his job.

Yes, Rex Berringer was good. If he knew people—and he knew people very well, lots of them—he'd have this case buttoned up before the day was done. He'd have Mister Dresden's list all filled out: What was ChronoCorp's mission? Why was ChronoCorp using so much juice from the grid? What were they not telling us?

Mr. Dresden was not your usual customer. Rex had encountered all types. Most kept to the shadows, or used disguised voices on the phone. They came in cloaks, raincoats, or vaguely menacing tailored black business suits. Mr. Dresden wore none of these things. He wore jeans. He faced Rex without preamble, and he was far more disturbing for it. He spoke with a southern drawl, and when he removed his expensive Gucci shades, his eyes sat within sunken pools of dark skin. He reminded Rex of a raccoon. He never doffed his expensive Stetson cowboy hat, incongruously white. Dresden usually chewed on something; whether it was gum or tobacco, Rex didn't know. Dresden kept his own background to himself, except for a slip about some experience with explosives and knives. The less Rex saw of Dresden, the better.

Melvin, DBA Rex, took the stairs down to the back lot and opened the door of his ancient Chevy van. There, he found his uniform and a magnetic sign that read "Ace Pest Control." He'd paid fifty bucks to a CSU Graphics student to design it, complete with a friendly insect logo. Another van was waiting at the rental place. He hated using his own for work. It always got scratched or dented or mangled, so he rented when his budget would allow. All he needed now was to stop by home, grab his silver-painted cardboard cylinder, complete with spray gun and plastic aquarium tubing (and filled with some sand to make it seem more substantial), and head out to the north of Fort Collins. ChronoCorp was hiding something. It was a part of the dark underbelly of lovely Fort Collins, and it would take someone special to bring it down. It would take a man of action. With a fake spray bottle.

*   *   *

The security guard worked his way down the corridor. This was not one of those hallowed hallways of Princeton. This building was a modern, boxy affair reserved for boring-sounding classes like "Geomorphology" and "Seismology." What was seismology, anyway? But like all the ivy-covered edifices, this one needed security, too.

A whole lot of humanity must have been in the Geosciences building today. He wondered why the Dean had been so insistent about his vigilance on the second floor tonight. Something about a new display? Exhibit?

After making the rounds on the ground floor, and making sure that all the basement doors were secured, he slogged his way up the east staircase. His dinner coffee was wearing off and he felt every step.

The second floor was darkened; the lights went off automatically at 10:30. He checked doorknobs and shined his flashlight through office windows. At the end of the hall, he turned west, expecting another set of darkened rooms. But toward the end of the hall, glowing through double doors to the specimen area, a blue light glimmered. It was faint, as if someone was sneaking around with a tiny flashlight. Maybe they were using their phone as a light source. Whatever they were doing, they weren't supposed to be here. It raised the hair on the back of his neck. He began to sweat.

He keyed his radio, called in a 402—possible intruder—and slinked down the corridor. He peered around the doorjamb. The room was utterly abandoned. The only light came from one end of a table. Across the table spread a bunch of old drawings lying under a sheet of plastic or glass. He stepped cautiously toward them. They were pictures of some kind of creatures, fanciful things, dragons or dinosaurs. He didn't know much about art. He didn't even know what he liked. But he knew something crazy was going on here. One of the drawings at the end of the table was giving off its own light. It had an eerie quality. He hoped he wouldn't be alone for long.

13

Just Following Orders
Wednesday, 7 pm

As Todd and Katya waited in front of ChronoCorp, Katya grabbed her phone. She peered over it at Todd. "I'm going to do a little background research on tonight's hostess."

"Good idea. Looking for something specific?"

"Just browsing." She spied Todd looking at her in a way he never had before. She fought back a smile and continued. "According to my favorite search engine, Mila van Dijk came into her fortune by some inheritance, a lot of hard work, and an advantageous settlement when her husband of two decades left her."

"Divorce can be lucrative for some."

"Seems like the cost is pretty high, no matter how you slice it."

"Yes, ChronoCorp knows all about her from their initial vetting, I'm sure. How did her ex come out of the whole thing?"

Katya swiped her phone several times. "Unknown. Doesn't mention a name. In fact, there's not much about her at all before she came on the philanthropy scene. She devoted some of her fortune to solar energy investments, private space exploration initiatives, and antiquities. She became famous for the latter."

"She sure did," Todd enthused. "That's where I first heard her name. Some story about saving Iraqi treasures from the war."

Their musings came to an abrupt end with Xavier's arrival. Katya stuffed the phone into her purse and stood at attention as the boss walked up.

* * *

M. Carroll, *Plato's Labyrinth*, Science and Fiction,
https://doi.org/10.1007/978-3-030-91709-8_13

Xavier was used to people following his orders. He'd been called a strong leader, and a talented manager. Still, he was pleasantly surprised by the fashion show unfolding before him. He couldn't remember ever seeing Todd in a shirt without some kind of phrase or logo emblazoned across the chest. But a collared silk with tie? And that was nothing compared to Katya. He expected her to show up in a modest business suit, perhaps an almost-matching pants-and-jacket combo over a nice blouse. Instead, she wore a gorgeous black evening gown off one shoulder. What he could see of her back revealed finely chiseled muscles, a product of all her before-work visits to the gym, he supposed. He remembered seeing a female weightlifter whose muscles he had considered unattractive. But Katya's reminded him of an athletic gazelle. Grace in strength. He wondered, briefly, how she got that zipper all the way up the back and thanked God he was born a gender that used buttons in the front. Though it was tailored and sleek, her dress was not designed to be provocative. No: this outfit projected power and confidence. It suited Katya, both physically and in personality.

Suddenly, he felt underdressed in his coat and slacks. He knew his pants were an inch too short; all his pants were. That's what he got for being 6′ 5″.

* * *

If it was Mila van Dijk's goal to keep them off balance, to impress and maybe even intimidate, she hit her mark. The "limo" was a chrome stretch Humvee, polished to a mirrored surface. The driver held open a double door as the three ducked into the spacious interior. He pointed out the bar and the controls for music and video, then took his seat behind the wheel. As the car pulled from the curb, Katya asked, "How long a drive is it?"

Xavier leaned forward, the leather squeaking beneath him. "'Bout twenty minutes."

"Our chauffer's speech made me wonder if we were gearing up for a week-long cruise."

Todd said, "You okay, Kat?"

"Somebody sure is trying to impress," she replied. "Isn't this all a bit…Saudi prince or something?"

"Yeah," Todd beamed. "Isn't it cool?"

They had a good-natured debate about what kind of music to put on. Todd wanted classical—preferably Debussy—while Katya preferred classic rock. Xavier leaned toward industrial. In the end, they rode in silence.

Todd fidgeted. Katya leaned over and patted his knee. "Is it about your trip tomorrow?"

"I shouldn't be so nervous. It's just a little farther than we've gone."

"You'll be fine," Xavier said. "This Q-slip will be a fascinating jaunt."

Katya put in, "How's the French coming?"

"Bon," he said feebly.

The street carried them to the west, through the Fort Collins suburbs, past the stadium and rows of apartments, up into the foothills beyond Horsetooth Reservoir. Their travel was in the direction opposite to Katya's neighborhood. Katya lived to the east of campus, in a big two-story house with a wide front porch, raised flowerbeds, a seventies-style avocado linoleum floor in the kitchen, and wooden shingle roof. Though it wasn't the most modern house, the place had character. She liked her house.

Mila van Dijk's house was nothing like that. The roadway split off into a wide, curving driveway that looped around on itself like a quantum equation. A twenty-foot-high fountain—what landscapers might call a "water feature"—babbled in three tiers of sculpted flowers, grapes, and cherubs. It looked very old, very European. Behind it loomed the van Dijk mansion, a sprawling three-story estate with columned doorways and Egyptian sphinxes lining the front walkway. Pink granite stonework framed plate glass windows, which mirrored the pine forests around them.

Katya liked Mila's house, too.

14

Visitation

At the back of a room illuminated only by a computer monitor, a conversation unfolded. A woman hunched over the screen, her features lit from below. She spoke into a hands-free set while she tapped at the keyboard. Then she paused and leaned back in her chair. Her shoulders stiffened. Into her headset, she said, "That's unfortunate. If your Cybersecurity people could have broken through, I might not have to—"

She hesitated for a moment, and then said, "Well, yes, I suppose that would be best…I am making some arrangements, but these things take time. After all, I don't normally do this kind of thing…no one suspects, I am sure. You can relax and trust m—"

She paused, letting out a long sigh. "Absolutely. I will see to it. But it's not going to be easy. I don't want to arouse susp—Yes, I know. Yes…naturally."

She tapped the earpiece and gently placed it on the desk with an annoyed groan.

* * *

Manicured hedges interspersed with lampposts bracketed the last half mile of road leading to the mansion. All the land belonged to Mila van Dijk. "No Parking" signs assured a clear run of rich green walls and faux marble bastions holding up gas-lamp-style lights. Next to one of the lampposts, a tiny man listing to one side trimmed the hedge. He looked up as the limo passed and waved, as if he knew they were coming. Katya thought the man should have been wearing a bright red gnome hat and big shoes.

© The Author(s), under exclusive license to Springer Nature Switzerland AG 2021
M. Carroll, *Plato's Labyrinth*, Science and Fiction,
https://doi.org/10.1007/978-3-030-91709-8_14

She expected a butler or footman or some *Downton Abbey*-looking servant to greet them at the door, but Mila van Dijk herself welcomed them. And she was welcoming. She stood just five feet tall. Her broad smile complemented an expansive, relaxed manner. Van Dijk's salt and pepper hair was coiffed, but with a small wayward tuft sweeping out the side. She wore a tailored suit, an ascot, and dark grey riding boots. No evening gown and stilettos here.

After Xavier made formal introductions, van Dijk stood in the center of the walkway, blocking their path. "Thank you so much for dining with me tonight! Before you come through, did you notice my sphinxes?" She jabbed a finger toward the front walk.

"They're beautiful," Todd said.

"Black onyx, Dr. Tanaka. A bit smaller than the sphinx we normally think of. Eighty-three centimeters at the shoulder. Just got them a couple months ago. A few are chipped, but for sculptures that have been slumbering under Sinai sands for two millennia, they've done pretty well. Got them from the Society of Egyptian Antiquities. All above board, of course. Special auction."

Katya found van Dijk's rich alto relaxing, almost hypnotic. "The money went to an organization that preserves and restores Egypt's national treasures. Some people think of such an investment as illegal, or at best immoral." Katya felt van Dijk's eyes on her, as if this bit were meant for her. "But that's foolish. My investment in these statues, the likes of which can be found by the dozen on the black market, funded preservation. Research. Advancement in archeology and even other sciences. Foolishness? Perhaps. I operate under the rule of *the greatest good for the greatest number*."

"Jeremy Bentham," Todd said.

"A student of philosophy," van Dijk murmured, perhaps really seeing Todd for the first time. "How refreshing. Of course, these carvings are new kids on the block compared to the timelines we'll be dealing with."

And there was the feeling again: van Dijk speaking to Katya, intimately, as if she knew her inner battle.

Van Dijk spun around toward the entry. "I like the way they glisten. Shall we have a few appetizers? This way."

Katya reeled. The piano nobile was a lot to take in: the ancient statuary, the Art Deco sweeping staircases, the jarringly modern Mondrian wall hangings, the vaulted ceilings. An original Alexander Calder mobile hung from the clerestory entry, bobbing slowly in a current undoubtedly engineered just for it. Hors d'oeuvres paraded across a room-length sideboard in the dining room. Just as Katya was wondering if van Dijk had any servants, two of them appeared, offering a plate of canapes and a before-dinner wine. Her appetite had evaporated, but she took a few items to be polite. Once everyone had a plate of goodies and a drink in hand, van Dijk stood at the head of the table and gathered them around.

"We will have much to speak of at dinner, so I wanted you all to have a look at my Minoan collection first. After all, that's what all your toil has been about. Dr. Stengel gave me a tour of your facilities a few months ago. I am returning the favor. This way, please. Feel free to bring your treats."

Katya couldn't help herself. Although their host was in tour guide mode, she wanted conversation with their patron. She wanted to know what made her tick. She wanted connection, not a soliloquy. As they made their way toward a large double door, she cleared her throat and said, "What did you think of ChronoCorp?"

Van Dijk paused, turned and drilled her with glacial blue eyes. "Impressive technology and provocative theories put to the test. I'm sure the National Science Foundation has been happy with the quantum research you've done, Big Bangs and black holes and such. I, of course, am more interested in your other research. Money well spent. Good stewardship. That excites me. I inherited some of this." She waved her hand down the hall and up toward the muralled ceiling, an immense Erte design looking down upon them. "But I took what I had and invested wisely. Stewardship. Husbanding what you have with care. I like what I see at your ChronoCorp."

Van Dijk turned back toward the double doors. She pulled one open as a dignified man in a suit opened the other. Inside, the long room was dimly lit, with cases illuminated from within. They had just stepped into a museum.

The first case held the finest Theran ceramic vessel Katya had ever seen. Never mind that she had only seen half a dozen in person; she had been studying photos of them for a year now, ever since van Dijk had come on board. This one was singular. It stood waist high, with double handles leading from a narrowed neck down to an oblong body. The ceramic was shaped like an amphora, except that it had a flattened base for standing on a surface.

Its shape was not what took her breath away. It was the painted patterns on its surface. A vine laden with purple grapes ringed the neck just below the lip of the vessel. Below it, a parade of ships sailed along, rows of oarsmen propelling them across dolphin-filled seas. Rosettes and spirals danced along the base. Much of the color was still vibrant, as bright as if it had been painted yesterday.

Katya took in a breath, close to a gasp.

"Yes," van Dijk nodded. "That's exactly what I did when it first came on the market. This particular pithos vessel was found among the contents of a market, buried under volcanic ash for nearly four millennia, part of the ruins in Akrotiri. It left Santorini in the late fifties, after the big earthquake, but it was repatriated to the Society, which in turn sold it at auction. They have a companion piece in even better preservation at the museum in Athens."

The tour continued with ever-finer treasures. Van Dijk stopped at a small waist-high case. Within floated a delicate necklace. "Jewelry is very rare on ancient Thera. Everyone apparently had plenty of warning, lots of time to leave and take their treasured belongings with them. But this…Nice, isn't it? Wouldn't you like to drape that one around your neck?" she asked Katya, caressing the surface of the clear case. "I like the beads on the lower register, each shaped like a Minoan double ax head. Look at the care, the craftsmanship." She stepped away as if leaving an old friend, and moved to a series of glassed-in wall pieces. "And now, to the city."

She brushed her hand across a switch inset into the wall. Soft lights washed over the surface. Before the guests spread a panorama of ancient buildings, a harbor, and terraced hillsides with vineyards. Van Dijk thrust out her hand dramatically. "I give you Thera, your destination."

Behind a pane of plexiglass, the cracked plaster had been meticulously reassembled from many tiny fragments. Some gaps had been filled in expertly with matching paint—guesswork for a mural painted 1,500 years before the life of Jesus of Nazareth. It seemed to Katya that there was a collection of Minoan artifacts that kept turning up in the literature over and over again. There were the boxing boys, the two fishermen, the House of the Ladies, blue monkeys, and of course, the famous Frieze of the Admiral's Fleet. Van Dijk's mural fragments could have been painted by the same artist, but the scene was unlike anything Katya had seen. Here were the familiar multi-storied houses, the great ships under sail, the festive harbors. But there was more. On one end, dolphins played alongside a small Minoan boat unlike the great oared ships of other murals. The ships seemed to be moored in a labyrinth of successive jetties, all rendered in the flat perspective so common in ancient art. Along the harbor, children frolicked with what looked like pet goats. One child had a monkey on her shoulder. In the background—something else not seen in other paintings—the cone of a volcano rose from the waters of the Aegean. The reds and ochers of buildings speckled its flanks, linked by zigzagging pathways etched into the green mountain. It was an idyllic scene recorded at a time when no one knew of the coming catastrophe.

"Mesmerizing," Xavier whispered.

Van Dijk nodded triumphantly. "Yes, the island that the ancients called *Strongyle* for its shape: round."

"And later," Xavier put in, "Kalliste, the fair one."

"Fair, indeed," she continued.

Todd leaned closer. "I wish the muralist had shown us a bit more, enough to solve the debate."

Mila's eyes flashed. "And what debate is that?"

"Nobody's sure just what the volcano in the center of Thera's inner bay really looked like. Was it a Fuji-style cone surrounded by water? Some suggest that the eruption was far too powerful for that. They say it must have been a supervolcano. If that's the case, the volcano itself may not have been an island at all, but rather some kind of land bridge across the outer ring of ground."

Mila grew silent for a moment. "Well, I suppose you'll be able to tell us that, in time. In this very mural, you see the genesis of Plato's Atlantis legend. He spoke of seven concentric harbors that incoming ships had to work their way through to get to the center."

"Sounds like rims of a caldera, concentric volcanic crater edges," Katya put in.

"That is indeed the thinking of some. A collapsed caldera." She pointed again. "Fountains powered by natural springs, and inside the homes they had hot and cold running water, along with toilets nearly four millennia before Thomas Crapper. Plato mentioned the sun shining off copper roofs, and stone buildings built of red and black masonry, just the colors we find in Santorini's volcanic layers." Mila dragged a finger along the protective glass, her eyes staring through to the mural as she followed it to the end. "It was Plato's Labyrinth," she murmured. As she reached the corner of the glass, she dropped her hand. She turned smartly toward them. "But of course my interest goes far beyond the Atlantis legend. I'm intrigued by the people, by the day to day living, the gymnastic celebrations in bull rings, the dancers and the fishermen."

Katya watched the burning glimmer in van Dijk's eyes, and wondered how exactly this priceless chunk of mural had fallen into the woman's possession.

"The fashion was exquisite," van Dijk went on. "The repeating patterns of labyrinthine designs, perhaps an echo of the famous labyrinth itself, at the heart of the Palace of Knossos just a hundred miles away. But you didn't have to be royalty to wear beautiful apparel. Look at those faces. Timeless."

Lovely soft features surrounded deep eyes and long lashes, with hoop earrings dangling across long necks. The girls and women wore their black hair in vertical curls corkscrewing down their cheeks. Full red lips looked as if about to speak. And what did that melodic speech sound like? What did these ancient queens have to say about life, politics, nature, the weather, the news of the day? If they got it right, Xavier's team would be able to answer those questions. And if Katya had her way, their work might save millions of people who came after that ancient civilization. A word here, a critical hint there, and who knew what advances might result?

Xavier had been more quiet than usual, but he broke in on Katya's thoughts now. "It's not that different from today, is it? All the beefcake and allure."

Surprised laughter erupted from everyone around him. He held up his hand.

"No, seriously. A lot of the men are shown shirtless in a working environment, or in some sort of games, while the women flow through the murals with their long skirts, their blouses open in the front. Perhaps all those open chemises reflected the openness of society?"

A server leaned in to van Dijk and spoke in hushed tones. The host said, "I think we'll wait on that for a little while. Check back after the meal."

"He says it is urgent," the servant said a little louder.

"Everything is urgent with him. Tell him I'm hosting a dinner. He's not invited, even by phone."

Van Dijk turned back to her guests. "You know, the ancient writings speak of the women and their music in the streets. But the texts aren't speaking of voices or lutes." She pulled a thimble-sized metal disk from her pocket. "An accessory from a Minoan woman's dress thirty-five hundred years past. They lined their hems with these, and the little copper disks would tinkle—sing, if you will—when the wearer walked down the street, as the sea breeze played with the fabric. Lovely thought, isn't it? Skirts singing in the ocean zephyrs? And wouldn't you love to smell the feasts they cooked, feel the cobbled roads beneath your feet, taste the smoke of the fires burning and the sulfur-scented fountains in the public baths?"

Katya's pulse raced. Van Dijk was describing a real place that Katya and her colleagues would soon visit.

Van Dijk's tone changed, no longer dreamlike, but now practical. "Now, I hope you can all see why I am so interested in researching this civilization. They were advanced beyond any peoples of the time in their control of the seas. The paintings and sculptures hint at the sophistication of their civilization, their social activities, celebrations, religious affairs, sport, mundane living. But so much was lost in the explosion." She closed her eyes and turned her face toward the ceiling before composing herself again to say, "and the tsunami. Pardon me if I repeat what you already know, but it is important that you understand my passion and the importance of your work. When that volcano blew its top, it covered the ancient Minoans in a fog that we have only partially cleared. After Thera exploded, the outlying settlements carried on, of course. But our main source of knowledge is not Thera—*Santorini*—itself, but rather the related provinces on the mainland and Crete. Historians would love to know what the hub of this civilization was like. What made Thera tick? I would love to know." She adopted a Mona Lisa smile. "Which brings us to your project. I have concerns."

"Concerns?" Xavier asked, the alarm in his voice not quite masked.

A woman stepped quietly into the room. Van Dijk paused and locked eyes with her. The woman nodded.

Van Dijk turned and gestured toward the door. "I believe we can continue this at table."

It was time for dinner at the van Dijk household.

15

The List

The rental van pulled up to the curb next to the Emerson Building, several floors of which hosted ChronoCorp. The sun hung low over the foothills, glimmering off the bronzed windows of the ten-story edifice. The lowest of the panes mirrored the street around them, including the vehicle with its grinning cockroach on the official-looking pest control sign.

Rex Berringer peered from the van window up to the crown of the building. Atop its formidable main structure, a somewhat mysterious tower rose another three or four stories. It looked like any other cell phone stack, but why have one on top of *that* building? All the other cell towers he had seen in Fort Collins were on the far side of town. What if it wasn't a cell tower at all? That was yet another thing he would need to find out.

Rex stepped jauntily from the van and dragged his fake pesticide sprayer to the front door. He wrestled it inside, wishing he hadn't put so much sand into it.

The directory said Suite 312. He took the elevator up to a dim hallway. The glassed-in reception area was dark. No office lights glowed down the corridor further in. Nobody was home, and breaking-and-entering wasn't part of his job description.

Rex hated loose ends. He despised not achieving his own personal goals for the day. He had a list laid out for himself each morning. The list always began with "Make bed" and "Brush teeth." It gave him a sense of accomplishment early on, even on those days when he got only one of the two. The momentum could carry him through life's little disappointments (like not finishing all the items on the day's list).

© The Author(s), under exclusive license to Springer Nature Switzerland AG 2021
M. Carroll, *Plato's Labyrinth*, Science and Fiction,
https://doi.org/10.1007/978-3-030-91709-8_15

He remembered a girlfriend—more of an acquaintance, really—telling him how life came in little packets or blobs, and each one was a treasure. Each one was valuable, to be cherished. Something like that. He couldn't really remember, because he had been preoccupied with planning something important for work.

The sad truth was that ChronoCorp would have to wait for another time, haunting him on the day's list as an unchecked box. What there was of his rational mind admonished him to not race home D.U.I.E. (driving under the influence of emotions), but the emotional component of his brain was screaming a different message: Why couldn't someone—*anyone*—have been there?

16

Main Course

The van Dijk dinner was indeed ready, and in lavish style. Fresh fruit, exotic and strange, formed mounds around an ice sculpture centerpiece. The frozen statue was not of a swan or a graceful figurine. Instead, what rose from the center of the table was a glistening, rotund White Rabbit from *Alice in Wonderland*, complete with a pocket watch. Its whimsical significance was lost on no one.

"So how do you wind that thing?" Katya asked with her best serious tone. Van Dijk raised an eyebrow.

Rainbows of color played through the chiseled ice and draped across the fruit at its base. Much of the produce must have been specially imported for the occasion; Katya hadn't seen it in the grocery stores she frequented.

"Oh, look; place cards!" Katya said, more energetically than she meant to.

"Very observant, Dr. Joshi."

Katya felt her cheeks flush. She began actively praying to whoever would listen that her ears didn't turn red. But they always did. Her high school chums used to call her 'stoplight.'

An attendant poured wine into crystal goblets as two others brought in potatoes au gratin and some sort of vegetable mélange. Katya heard the main course before she saw it. A servant had entered at the back of the dining room with a sizzling plate of still-broiling salmon. The pink fish—the largest filets Katya had ever seen—lay beneath a forest of fresh spices and a carefully placed mandala of sliced lemons.

"Well, Dr. Stengel, I have told you why I want you to go. Now, why don't you tell me why *you* want to? Why play with the dice of the universe? I know you well enough to realize that your motivation is not money." She leaned

M. Carroll, *Plato's Labyrinth*, Science and Fiction,
https://doi.org/10.1007/978-3-030-91709-8_16

toward him and tapped her finger in front of his plate with a subtle smirk. "So tell me, what is it that banks the fires of your soul?"

Katya felt as if she was eavesdropping upon a very private conversation. Xavier stared at his generous meal and put his elbows on the table. He steepled his fingers for a moment, then looked up and held Mila's gaze.

"When you're a kid, 'out there' seems enigmatic and magical. It's the hinterlands. You know your neighborhood, but there's enchantment beyond those paths you usually walk, the roads your family car travels. For most adults, those territories out there—the ones inscribed with fanciful sea monsters and pirate treasure—become predictable. Out there in that mysterious no-man's-land are only farms; the abandoned mill is just an empty wooden shell; those distant green hills are just more of the same cow pastures." He leaned in toward the table, a gleam in his eyes that wasn't there before. "But time travel has filled us with the wonder of childhood again. It's transformed 'out there' into a magical realm once more. Who gave us the heart to dream these dreams? Wells gave us the taste for it. Einstein and Heisenberg and Mallett were the chasers, giving us the wherewithal to make it happen."

"And—speaking as a businesswoman and investor—you see the promise of more to this research?"

"We don't know all the possibilities, of course. Not yet, anyway. But we know there is potential." He leaned forward. "Tremendous potential."

Katya could sense the tension rising. Someone had turned up the gravity in the room. Xavier was ramping up again, lapsing into his thespian mode, and all she could do was sit back, marvel, and hold on for the ride.

"As I think I told you in our first talk, breaking the time barrier was never what we set out to do. We were simply trying to unravel a few mysteries about the quantum universe. Our quantum slips came as a side effect of other work. But our Q-slips are more than traveling to another place and time: they're destiny with a capital 'D'." His hands were working now, arms moving like an orchestra virtuoso conducting *Flight of the Bumblebee*.

"And it's far more than trips to nineteenth-century New York or ancient Thera, as valuable as they are. Here we have the key that opens a vault, a black box that was, until now, inaccessible to quantum physicists and theoreticians."

"Origin of the universe?" van Dijk offered.

"Exactly right. And more, undoubtedly," Todd said. "How does it work? What dark energy is driving it apart? What, really, is dark matter? And theory, eventually, becomes practical. What we find behind that locked door may change the very way we all live."

"It already has," Xavier said. "This kind of physics has given us lasers. Transistors. Cell phones. The quantum universe offers unimaginable advances in computing, materials, flexible screens, nanorobots. It most certainly will change our view of reality, of how the quantum cosmos works today and how it came to be."

"You make it sound very…clinical," van Dijk said, baiting him.

"But it's not clinical at all," he objected. "It's the essence of adventure. It comes up and confronts us. We humans are creatures of thought, of ideas. We were formed as substance and raw energy, but we've transformed ourselves into inspiration and resolve. And that resolve propels us—*propels us*—to create and to explore and to go beyond!" Suddenly, a look of embarrassment shadowed Xavier's face. Quietly, he said, "That's how I see it."

Mila tore a dinner roll in shards and addressed the others. "There, you see? This is why I wanted to fund your research group. There are others working the issue, but Xavier's enthusiasm matches mine."

Katya had been observing, holding back to watch her host. Now, she held up a finger and said, "You mentioned having some concerns?"

Xavier shot her a death stare just brief enough for van Dijk to miss.

"Yes, I suppose now is a good time." Van Dijk tumbled some expertly broiled vegetables onto her plate. "I am bothered by some of the logic. After all, what you are doing appears to fly in the face of rationality, of the way nature unfolds. I do action 'A', and action 'B' follows. But what you've done is to make it possible for B to precede A. It's backwards causation. It violates fundamental laws of the natural order."

Katya swallowed hard and said, "That's relativity, not quantum mechanics. Relativity is about space and time, while quantum mechanics is about the universe on the atomic and subatomic scales."

Van Dijk foundered, apparently at a loss as to why the distinction was important, so she said, "It's the old story of going back to kill my grandfather. Will I still be here? Will I sprout another universe?"

"Multiple universes solve some of the problems of paradoxes," Todd said, but van Dijk was on a dinner roll that wouldn't stop

"The multiverse," van Dijk scoffed. "You do as you will, without care or accountability, and when you die you simply wake up again in another reality?"

"That's only one version," Todd protested. He opened his mouth, but that was as far as he got before van Dijk was in full swing.

"If there are parallel universes, you could take advantage of that, too. An unscrupulous man might send himself just a little into the future, barely offset from the reality we know, and wait patiently. Then he would have a twin appearing in the ranks. He could repeat the process ad infinitum and create a loyal army, or just a second person to do some dirty little deed."

"I can see that," Katya said. "A second universe buds off this one, but if the time difference is slight, the two might be able to interact."

Todd held up a knife loaded with butter, slapped it on a bun, and said, "Actually, it's not that far-fetched. There may be some evidence hinting at a few places where other universes bump into ours. If you sent someone a short enough throw, say, a few minutes into the future, their universe and yours might coincide. Coexist. Something like that."

Xavier put his fork down on the table. His hand quaked subtly. It wasn't like him to be nervous in such discussions; he embraced them. Katya wondered if he was okay, but Xavier forged on. "Perhaps. But this is where you're wrong, Mila. First, some models say we can't go forward in the timeline. No future travel. Maybe it's just a matter of solving some engineering problems or looking at quantum equations in a different way. But some of our colleagues liken the movement through time to a spiral that goes backwards, beginning from its starting point in the present. It doesn't go forward, and it doesn't go directly backward either. It's essentially a corkscrew through space/time. As for going back in the progression of time, it's not that you can't go back into the past. It's just that you can't change it."

Van Dijk squinted at Xavier. Was she having second thoughts about her funding? They were all committed now, up to their eyeballs in new technology. Xavier seized the gauntlet. "There's a brilliant time expert named David Lewis. Have you read his books?"

She shook her head.

"He claims that all these apparent paradoxes are oddities but they're not show-stoppers. I think he's right, and it's not just that I'm dying to see those Minoan fleets of yours." He pointed a bread stick in the direction of van Dijk's vault.

Van Dijk hooked a finger, nonverbally calling for one of the servants. "Shall we have dessert?"

"Certainly, madam."

"You'll all like our little after-dinner menu," she told her guests. "And yes, I must say, what you're doing is exciting and filled with great possibilities, as you say. But like all technology, it can be misused. With power comes responsibility. And I've been paying your energy bills. You use a lot of voltage. What could be done with all that power in the wrong hands? Or better yet, what could go wrong? What are the probable—or even improbable—consequences?"

Xavier started nodding before she finished the sentence. "The Jurassic Park syndrome. With enough variables, what can go wrong will go wrong."

"Murphy's Law," Todd put in.

"Yeah," Katya said to Todd, "but with a whole lot higher stakes."

"I'd say a pack of Velociraptors was high enough stakes, wouldn't you?"

Katya spoke around some scrumptious morsel. "Mmm, Velociraptor steaks. Wonder what those taste like."

"All good points," Xavier said, trying to take control of the discussion. He reconsidered, glancing at Katya. "Well, some points were better than others. But at any rate, we think time has a way of healing itself. That's why the paradoxes are non-starters. If I go back in time and kill my grandfather, by design or through accident, the universe I return to will have mended that action in such a way that it would simply disappear. Perhaps he can't die, or perhaps any weapon I use will be ineffectual. Or maybe I can't physically get to him. Either way, all is right with the world."

"Even with all those ripples in the pond you're making? Spreading out and out? What do you think, Ms. Joshi?"

Katya was caught off guard, but she was used to fielding questions from the press and the skeptics. In fact, Xavier had appointed her as a de facto press liaison in the past, before they entered into more secretive work. "That analogy is part of the problem. You see, time may not really be like a pond. A better image may be a river. In a quiet pond, you drop a pebble and the waves expand out, bigger and bigger circles. But what happens in a river? The pebble makes a little splash, the ripples go along for a bit, but they get lost in the currents of the stream. They disappear into the background noise. All returns to normal."

"And what if you try to make big waves in that river?"

Katya shrugged. "The flow of time may even them out. As Xavier put it, self-healing."

"This is a great deal of guesswork, of course," Xavier put in.

"Which is why you are doing the research. And as for misuse of our fine technology?" van Dijk prodded.

Mila was looking at Katya, but Xavier intercepted. "I think we can all be thankful that you've vetted everyone on the project. You have not detected any evil super-schemes, I assume. We're all too busy to hatch any extracurricular activities with our time pods, rest assured."

Van Dijk laughed. Katya thought the sound had a forced edge to it.

17

Luxury Vehicles

As they left the mansion in style, relaxing in the limousine reserved just for the three of them, the ChronoCorp trio spotted the gnome-like gardener again. The failing glow of twilight revealed him sculpting a shrub menagerie at the side of the main property, lit by ornate lamp posts. He gave them another nod. The other support staff remained fashionably invisible, except for the attendant who had helped them through the spacious doors of the limousine. Finally, Katya thought: someone who *does* look like they belong on *Downton Abbey*.

The ride down the hill was boosted by robust mugs of fresh-ground coffee, an exotic concoction van Dijk ordered specially from a tropical country with lots of rain and nasty bugs. The trio of guests even got to take their mugs home with them. Katya wondered if all of van Dijk's guests got coffee cups. Or maybe tee shirts.

She leaned over and taunted Xavier, who was yawning like a lion. "Now I know why you want to go back. It's that Santorini fashion."

"Not me. I've been to the beaches of France. After you've seen a few topless swimsuits, it gets to be background noise. Besides, if anybody in the office is a pervert, it's Todd."

Todd glared at Katya. "Well? Aren't you going to defend me?"

"I'm thinking. I'm thinking."

And she was, but she wasn't thinking about bare-chested Minoan princesses or Todd's moral compass. Her mind was on Mila van Dijk's comment. *What could go wrong? What are the possible consequences?*

And Xavier's retort about a self-healing universe. He was probably right. It did, after all, solve those pesky paradoxes. But his response had seemed empty

© The Author(s), under exclusive license to Springer Nature Switzerland AG 2021
M. Carroll, *Plato's Labyrinth*, Science and Fiction,
https://doi.org/10.1007/978-3-030-91709-8_17

to her, a diversionary tactic. What did he really think? She turned back to him, but he was snoring. She let him. The smooth ride of the limo lulled Katya, too.

Suddenly, the vehicle lurched. She heard a horn blast just outside her window and looked up in time to see the flash of a white van. On its side, the countenance of a smiling cockroach bore its teeth at her (did cockroaches even have teeth?). A banner above the bug declared, "Ace Pest Control."

"That van is the pest," she muttered, and closed her eyes again.

* * *

Delaney called across the cluttered room. "Dr. Glenn…paging Dr. Glenn. Line two, please."

Brad picked up the phone on his desk, shaking his head at her. "Always so entertaining."

She said, "Chad Markham of Princeton?"

He punched a button and said, "Paleontology. This is Brad."

He exchanged pleasantries, but they were over quickly. Brad paused for some time. A slow frown washed across his face.

"I see."

Delaney watched.

"Yes, I've done some research on—no, no, nothing like that. More of a hobby."

He had his humble look on, but became serious. "Well, yes, that was my paper. Say, what's this about, anyway?"

Ivan stepped into the room. Delaney raised a finger to her lips as Brad said, "Well, I'm sure that's quite a find. Could be historic."

Who found what? Delaney wanted to scream. She and Ivan began to hang on every word.

"I see. Well, I suppose that would mean—But isn't it possible the drawing is a modern fabrication of some kind, some forgery left by…Ah, yes, well. And buried in between all those old papers."

Brad was furiously writing something on a pad now.

"In a crate? That's remarkable. Yes, right away. I can be there as early as end of the week. Yes. I'll get details after I chat with my boss. Not at all; I'm happy to help out. Thanks for the call."

Brad looked up. Both Delaney and Ivan stood silently, slack-jawed. He couldn't help but laugh. "Ivan, I think I'm going to need to take a day or two off for a little visit to Princeton."

18

New Digs
Como Bluff, Wyoming

It only took Ajit a few hours to settle into the cadence of the encampment. The vitality of his grads gave him a new energy of his own. Not all of the student diggers were paleontology or geology students. His conscripts represented many sciences and a few other disciplines, as Rodney Giles so aptly demonstrated. Rod's major was art. He was digging in the layers of Como, like many, to pick up a few summer credits. But in the shade of an afternoon break beneath the awnings, the student was studying his phone.

"You getting reception on that thing?" Ajit asked.

The student looked up, his mop of black hair dangling into his eyes. "No such luck. I just downloaded a few news stories this morning in Bow. Don't know why I bother, though. It's pretty much all bad news."

"Like what?" asked a sunburned blonde woman. "I've been cut off from any social media for…"

"Hours?" Ajit grinned.

Rod studied his phone again. "War, pestilence, upsetting sports scores. You know how it goes."

The woman grad sighed, slathering more sunblock across her forehead. "It's sad. There are better ways of making a new world."

Ajit said, "Other ways take patience."

Rod tried to swipe his screen to other subjects, but it remained in digital limbo. He glanced over his shoulder at Ajit. "I love how the things you do and what I do dovetail."

Ajit sat down out of the sun and glanced at the little screen. "Dovetail?"

M. Carroll, *Plato's Labyrinth*, Science and Fiction,
https://doi.org/10.1007/978-3-030-91709-8_18

"You know, art and science. Lots of the biggies kept fine sketchbooks. Marsh and Cope, Edward Hitchcock, The Mantells. And of course there was Charles Knight, the Michelangelo of Mesozoic murals."

"Ooh, Mesozoic Michelangelo," said the other grad. "Sure your major's not poetry?"

Ajit knew these things, of course. But he pressed the young man, "Do you suppose with things like our cellphone cameras, all that's becoming a lost art, so to speak?"

"Mm." Rod had something on his mind. Ajit had seen it before: the reticence to correct one's superior professor.

"Ah," Ajit encouraged, "but you don't think so, right?"

Rod dabbed moisture from his forehead. "Well, today we've got our own artists, the Robert Bakkers and Eleanor Kishes. Hallett and Sibbick, Luis Rey. Lots of others."

"And now they've found those old drawings in New York," the sunburned scholar added.

That got Ajit's attention. He gulped the last of his water and gurgled, "Do tell."

She held out another water bottle to Ajit, just as he set down his empty. He appreciated her kindness. Her name was Crystal, he thought. He had to memorize this crop of student names before the end of season. She was a smart one. Lots of potential there. Some people had a natural gut reaction to the rocks, an intuition about what lay within the earth beneath their feet. She was one of those.

Crystal said, "You didn't see the newsfeed about Princeton? Apparently some drawings turned up by Waterhouse Steven Hawkings."

"Hawkins," Ajit corrected. "Benjamin Waterhouse Hawkins." That was odd. First his sculptures unearthed in Central Park, and now this, all within a couple months.

Rod went on. "Yes, yes, they're causing quite the stir among collectors and historians and such."

Ajit eyed Rod's phone skeptically. "Sure you don't get reception on that thing up here?"

"Not a drop."

Rod was absolutely right. Art and paleontology. The overlap of disciplines. And two discoveries at once? It was too much of a coincidence: he suspected he knew what Kat's first destination had been. Apparently she had made it there and back again, just like a little hobbit. Ajit was already on his feet. "Tell Marta I'll be back in a day or two. She's in charge until then." He scampered up the ridge.

"But you just got here!" Rod called after him. Ajit paused and turned. "I might be considerably longer, but I'll try to get word back to camp."

*   *   *

Ajit hated off-roading, but even the best wilderness pro would have been proud of his drive back to the highway. They had just been talking about time travel over burgers, and now these new drawings appear from 1880? No. This was no fluke. It must have something to do with Kat, and the answer, he suspected, lay at the feet of Waterhouse Hawkins.

His cell was sporadic at best in Medicine Bow, so he set out for the Virginian, the historic centerpiece of the town. The hotel's landline would be more reliable. This time, somebody was on the other end. The voice was from a woman he didn't recognize.

"AMNS Paleontology. This is Delaney."

"Hello, Delaney," Ajit said in his most professorial voice. "This is Dr. Ajit Joshi. I'm trying to get in touch with Bradley Glenn."

"You've come to the right place," she said brightly. "Hold one moment, please."

Ajit could hear Delaney hold the phone away. Or perhaps she put in down on a desk. After some static, Delaney's voice came through clearly. "Brad! Get your butt in here. You have another admirer on the line."

Mumbling ensued, from which Ajit could hear the phrase "not very professional for a museum of this stature," and something else about shoes. A man's voice answered.

"Bradley Glenn."

"Brad, Ajit Joshi."

"Ajit! I was trying to track you down. I even called your daughter. I thought you were off in the Wyoming hinterlands."

"I was until half an hour ago. I'm calling from Bow. So you and Kat talked?"

"I wouldn't put it that way. We had a very brief chat about how I might get in touch with you. Not very helpful."

"She wasn't?" Ajit barked.

"I mean the conversation was not very helpful. I'm sure she's a wonderful girl. Daughter. Person."

Ajit was not convinced by his tone, but he had to get to the business at hand. He said, "It's about those drawings they found at Princeton. Have you seen them?"

"Actually, they've called me in to take a look. Damnedest thing, right on the heels of finding those statues in the Park."

"Yeah, that's what I was thinking."

"I'm suddenly in demand; instant celebrity, all because I got interested in some obscure British sculptor."

"You're definitely the go-to guy on this. How does it feel to be famous?"

"I wouldn't wish celebrity on anyone. I like my quiet. And this is anything but quiet. Not very entertaining, really. Want to come meet? See the treasures for yourself? I'd love a second opinion."

"Love to. How soon?"

"Soon. I'm driving down day after tomorrow."

"I could get a flight by then. Mind if I bring someone along?"

"I guess that depends."

"I was thinking Kat would love to see these."

The line became very quiet for moments. "I guess if it's okay with her, sure. What's her tie-in?"

Why wouldn't it be okay with her? Ajit wondered. He said, "The Hawkins renderings may parallel some research she's doing. In a different field, but perhaps related."

"You know I'm not looking for some relationship at the moment. Once burned—"

"Don't be so suspicious. I'll see if she's available. For travel, that is. We can take a car direct from the airport if the flight times work."

"I'll put in a word with a colleague at Princeton. They say it's a circus. Reporters and dinosaur fans and curious rubberneckers."

"It does strike me as imprudent that they let the story out so soon after finding the things."

"There wasn't any official release. The story was forced out. The student paper sent reporters to see what was up, things got leaked, and before anybody could do anything, it was in the mainstream public news feeds. Online *Boston Globe* and *Orlando Sentinel* were the first—don't ask me why. And I'll tell you something weird: One of the drawings isn't like the others. According to my buddy Chad Markham, it's of an Iguanodon. You know Hawkins' reconstructions of the Iguanodons, right?"

"Saw one of the sculptures at the Crystal Palace when Marielle and I were in London. Lumbering lizard, built like an iguana with a thyroid problem. Clawed feet, dragging its scaly tail, pretty classic stance for those days."

"Typical of early dinosaur restorations, right? But there's one drawing that has the beast streamlined, with its spine nearly parallel to the ground, tail out like an English Pointer, belly up off the ground, forearms small and agile."

"Not your standard nineteenth century fare," Ajit admitted.

"And across the drawing was a notation that said the image was not enough like a lizard. Something to that effect."

"Puzzling."

"Damned straight, it's puzzling," Glenn nearly hollered. "Something strange is going on and I want to see the thing for myself. All I can figure is that the one drawing is a fake. Must be."

"The other possibility, that Waterhouse Hawkins had somehow intuited the real stance and features of a creature that, at the time, was only known from a tooth and a thumb spike, is astounding."

It was as if someone had tipped Hawkins off, and Ajit knew who that someone had to be. He had to see the rendering.

"I'll let you know as soon as we have tickets," Ajit said. He hung up. He had to book airfare and call Katya, and not in that order.

*   *   *

"They call it a 'remarkable coincidence,'" Todd read from his screen. Kat and Xavier stood looking over Todd's shoulder at the news brief.

Katya leaned in closer. "Remarkable is right."

"But coincidence?" Todd stuck out his lower lip.

"I don't believe in coincidence," Xavier said, popping a white tablet into his mouth.

Katya put a hand on his shoulder. "Still got the headaches?"

He nodded, eyes closed. His Adam's apple bobbed up and down for a moment, and then he looked at both of them. "Guys, as for those Hawkins paintings, and the sculptures, perhaps it's not so puzzling. Time seems to move in courses and eddies and countercurrents. It links and entangles. The ancients called it synchronicity. Whatever it is, I don't think it's any coincidence that Princeton discovered Hawkins' drawings at the same time somebody was digging up his lost sculptures in another state."

Todd scoffed. "So you think on some cosmic level it's all tied in?"

"Consider the other options. Kat visits nineteenth-century Central Park, has her chat with Roundhouse Hawkins—"

"Waterhouse," Todd and Kat said in unison.

"Yes, well, *Hawkins*, shows him a few images against our rules of engagement—"

"I said I was sorry," she grumbled.

"—and then, lo and behold, we see this drawing of a new-looking dinosaur in a very old-looking crate. No, I don't think it's coincidence at all. What bothers me is that if her little meeting really did change the course of his work, something lasted beyond the quantum slip."

"Something stuck," Todd said, looking at Katya. She was looking at the floor.

"So what does this mean?" Xavier asked the air around him. "Does this mean time *can* be changed?"

Katya looked up hopefully. "Could the drawing be a forgery?"

"To what end?" Xavier sounded unconvinced. "Who would gain? One thing is certain: before we do much more in the way of running back through history, we need to examine those drawings."

The phone rang. It was Ajit Joshi.

19

Short Day

In the end, ChronoCorp shelled out for two tickets to New Jersey. Xavier and Kat would be packing that evening, flying the next morning. But just before noon, Ajit called Kat again. She put him on speaker.

"My dear, I had a strange chat with Professor Markham at Princeton."

"The guy in charge of the drawings?"

"Well, he's in charge of the department. When I gave him our flight times, he told me to hurry."

"What's the big rush?"

"Apparently, the drawing in question has some strange pits all over it, and they're spreading. The Princeton conservator can't figure it out, and she hasn't been able to stop the process."

"What about the other drawings?"

"They don't seem to be affected. Fingers crossed on that, though. If it's some kind of mold or something, it could wipe out the lot." After a pause, he said, "You're sure you had nothing to do with this?"

She looked at Xavier, then at Todd, and then at the phone. "I'm sure," she lied. She would make it up to him later.

Katya signed off. She looked at the others. "It's no mold."

Silence settled over the room. Todd cleared his throat. "Maybe this is that self-healing thing Xavier's been talking about."

Xavier scowled, his mood clearly darkening. "Still, the implication here is that Kat changed something in the past that had a lasting effect." He slammed his palm on the table with a loud thud. "All travel is cancelled until we understand what has happened, and I'm not talking about airline tickets." He sat heavily in one of the plush conference room chairs.

© The Author(s), under exclusive license to Springer Nature Switzerland AG 2021
M. Carroll, *Plato's Labyrinth*, Science and Fiction,
https://doi.org/10.1007/978-3-030-91709-8_19

Katya felt like screaming. It was beyond frustrating: plans for the Thera Q-slips were just about finalized. But perhaps this was the price they would have to pay for their new gift to humanity, the back and forth of knowledge, the infusion of progress from future to past and back, with its impact on the present again. "We'll figure it out," she said, banishing the doubt from her tone.

Todd nudged Kat's shoulder. "Hey, do you want to grab lunch after work? Before your big trip tomorrow?" He wiggled his eyebrows meaningfully.

"Um, yes?" Was that the right answer?

"Let's try something different. A college buddy of mine told me about a new Greek place south of town. See you in a few."

"We should all get a good night's sleep," Xavier said. "For tomorrow."

"Good thing it's Wednesday," Todd said a little too cheerfully. "Of course, I love every minute here, but half-day is the high point of the week for *some* people."

"Yep," called the janitor from out in the hall. "I'm off to the Ram's game. See you all tomorrow."

"I think I'll give everybody a short day," Xavier said, massaging his temples. "I know I could use one."

One by one, the lights of ChronoCorp blinked off at lunchtime. The company had a small contingent of security people, but most were the kind that searched for software breaches. Today, a lone guard patrolled the halls and monitored the front lobby, checking in at the CCTV cubicle periodically.

Xavier took his headache home to a cold compress. Half-day.

20

Roaches and Restaurants

Sadly for Rex Berringer, he'd had to return his nondescript white van to the rental company days before. It had been perfect for the job, but today they had none on offer. In fact, the only vehicle available was a fairly late-model Volkswagen Beetle. It was lime green. That would work. Rex slapped the "Ace Pest Control" sign onto the side of the car—it was almost too big for the door—and headed across town to the Emerson Building. He dragged his faux bug sprayer to the elevator, punched Floor Three, and put his sunglasses on for effect. The doors opened to a strangely dim hallway. He pulled his shades off. It wasn't him: still dark. A person in a janitorial uniform stepped through the reception's glass door and turned around. He was locking the place up.

"What gives?" Rex whined. "It's the middle of the friggin' day. Where'd everybody go?"

The janitor jangled the keys as he stowed them on his belt. "Short day for everybody."

"On a Wednesday?"

The employee shrugged dramatically. "At least we don't have to wear Hawai'ian shirts. Office opens tomorrow at eight-thirty, but someone's usually here earlier. If it's urgent." He was studying Rex's cylinder doubtfully.

"I can't even believe this," Rex mumbled. To the janitor, he said forcefully, "So glad you all got a day off. Wish I did."

With that, he left. He tried to storm down the elevator, but elevators go at their own pace. Plopping into the VW, he didn't even think about self-disciplined driving.

© The Author(s), under exclusive license to Springer Nature Switzerland AG 2021
M. Carroll, *Plato's Labyrinth*, Science and Fiction,
https://doi.org/10.1007/978-3-030-91709-8_20

* * *

Todd's suggestion for Greek cuisine was fine with Katya. She'd had her fill of burgers and "Microwave Masterpiece Insti-Chow" over the past weeks. Café Crete was new, and promised a refreshing change. Her mind was on the restaurant instead of the road when a day-glo green Volkswagon careened through a stop sign and swung across the road in front of her, weaving into position in the oncoming lane. The tiny auto rocked back and forth as it passed her. The driver leaned on his horn as he rocketed by. Katya caught a glimpse of a colorful sign on the side of the vehicle. She had seen it before, but she couldn't quite place it. She took a deep breath, willed her pulse to slow, and pulled into the parking lot.

Todd was waiting in a booth when Katya stepped through the front door. She expected the décor to drip with kitsch, but instead, a tastefully painted mural of a fishing village spanned the back wall, accented by indoor tables with umbrellas and candles. The painting reminded her of Mila van Dijk's Theran murals, while the tables reminded her of an outdoor restaurant she had visited in Italy. The effect was soothing.

She started talking before her butt hit the cushion. "So you did that up-and-down thing with your eyebrows when you said we should come here." A little tray on the table held slices of feta, black olives and onion spears. She reached for an olive.

"What up-and-down thing?" he asked with a tinge of defensiveness.

"That up-and-down thing where your eyebrows look like caterpillars doing the samba all over your forehead. I mean, I'm flattered by your invitation, of course. But you always do that eyebrow thing when something's up."

He grinned.

She said, "So, what's up?"

Todd started in as soon as the waiter dropped off the menus. "I've been thinking."

"I know that's hard for you. About?"

He stirred sugar into his iced tea, but his eyes were on her. "We've been so busy looking at how to do the science that we've forgotten the why. What do you think? Have we gone too far?"

Kat still had an olive in her mouth, but she stopped chewing and stared at him. "I was thinking we all had a pretty good handle on things. Are you thinking we've lost track of what's behind all this? Of where it's going? Seems like we have those discussions fairly often. Maybe you're just projecting." She said the last playfully.

"Something like that. But I got to thinking that maybe it's simpler than people think. Maybe it all comes down to free will versus fate."

"Sounds awfully deep for a Greek restaurant conversation, but then you're always bringing up those deep questions. This time I'm afraid I have no idea what you're talking about."

"No, not so deep," Todd objected amicably. "It's like this: If the universe wants us to break the quantum barrier, to slip back in time and change things, it will let us. In fact, it has already dictated that we do it. It's a done deal. We don't need to worry about it, because God or, if you prefer, the universe or fate, has already declared it to be so. The question's too big for us to decide, so it's decided for us."

"You're sounding a lot like my Catholic grandmother right now."

"In perspective more than in practice. But our friends with a religious bent might be on to something. Maybe the Calvinists were closer. They suggested that those destined for Heaven are predetermined, but us poor joes down here on Earth should treat everyone as though they are chosen children of God. No prejudice."

"I like that."

"Me, too."

ChronoCorp's own friar, Kat thought. How appropriate that he would be running off to medieval France if all went well. He would fit right in.

The waiter brought Greek coffee in individual carafes, each with its own little Sterno burner. The sweet concoction gave off an aroma that upped Katya's beats-per-minute even before it passed her lips. Todd pushed his iced tea out of the way and held his small coffee cup aloft. "This stuff could hold up a spoon."

"It could melt a spoon."

Todd smiled, but it faded quickly.

Kat took a sip, swallowed hard, felt the burn of the brew as it made its way down her throat. The aroma made it back up into her nose. She could almost smell the olive fields and vineyards of Santorini, the salt air, hear the creak of the windmills and the scream of the gulls. She pulled herself back to Fort Collins.

Todd wheezed. "We should bring this home for Emil. He could use it to disinfect the lab."

Katya nodded. "Dissolve everything off the counters." She put the cup down, studied it for a moment, and said, "Don't get me wrong: cosmology and physics and astrophysics and all the other sciences, all that makes me think of the universe as having some sort of organizing force—that's difficult to deny. Besides, I've done the math." A smile crossed her face.

"Hey, it applies. If John Calvin had our pods, I'll bet he would say, 'Do as you will. It's all been predestined anyway, by Someone much wiser than you with a whole lot of his own universes.'"

"And I would think you of all people wouldn't like the concept of a multiverse. It seems like a fine bit of evidence for a secular view."

"Were you thinking that because there's more than one universe, there's no room—no reason—for God? Maybe your God is a little too small." He said it with a glint in his eye. "Given that there are universes out there coming and going, appearing out of quantum probability, only to collapse, popping out of existence because they're unstable…then you come to ours. All those others are transient because of their makeup. Too much dark matter, bad ratio of weak to strong atomic forces, the right speed of expansion, on and on. Ours has all those things in just the right quantities. It's as if it's been fine tuned."

"Or luck of the draw, and there may be a whole lot of drawing going on out there." Katya wrinkled her upper lip in mock disdain.

Before she could give him a comeback, he asked, "Luck? Design?"

A young waiter had approached silently, but now he cleared his throat. How long had he been waiting? Hurriedly, the two of them ordered. When the waiter left, Todd dropped his voice, glanced around to make sure no one was still watching, and said, "But that's not why I needed to talk to you in private."

"Didn't think so."

Todd straightened in the seat. "Xavier's been on more trips than anyone."

That was no secret. "Riiight. And?"

Todd leaned forward. "Have you watched him lately? He's got the shakes or something. Sometimes I feel like he's on the verge of keeling over. Just collapsing in a pile of gyrating jelly on the floor."

"Yeah, I get that impression, too. He's on edge."

"No, it's more than that. It's like he's about to lose it completely."

"Do you think the boss is sick?"

Todd looked out the front window for a moment, and then met Katya's eyes. "Did you get headaches from your Q-slips?"

She nodded. "Some. And vibrant nightmares. But I've never been prone to headaches, so that was weird."

"I got headaches, especially after my second trip. And all our trips have been short sorties. What's it going to feel like when we've been gone awhile? That last one was enough time to give me a major migraine. And Xavier gets them, too. They seem to get worse the more you travel. But I haven't gone back for a while, and Xavier has, and he's miserable. You can see how every Q-slip takes a little more out of him. More and more. In fact, I think he's gone back more than he's saying."

Now, Katya leaned forward. "I've been wondering about that. There are gaps in the primary log sheet, after-hours lines where there are blanks instead of routine entries."

"That's not all. I looked up our power curves. We've had at least three prolonged spikes that Emil can't explain. But if Xavier's gone out..."

"Why wouldn't he tell us?" Katya interrupted.

"Who knows? Maybe he's not sure it's all working as it should. He's very paternal when it comes to us baby chicks, you know. Or maybe he was trying out the latest imager."

"If he's been going out, without us to monitor, it bothers me that he's being so secretive about it. It's a little..."

"Reckless?" Todd filled in.

"Risky. Slapdash at the very least."

Todd rubbed the back of his neck. "Truth be told, I haven't been feeling so great myself."   "Should we confront him on it? Do some kind of intervention?"

"That would be awkward. He's the boss."

They both sat quietly for a few moments. Finally, Katya broke the silence. "I guess we don't need to say anything unless things get out of hand."

"How do we know when that is?"

"I have a feeling it'll be obvious. I just hope he does okay on our trip tomorrow. Airport security is enough to put anyone over the edge. Too bad we can't zap ourselves a couple days into the future." She frowned. "I wonder about travel in the other direction."

"Into the future?"

"That worries me almost as much as going back. Of course, there are fewer logical issues with paradoxes and such. If I go forward, my actions don't have the same kind of ramifications as going backwards."

Todd dropped his chin and spoke into his coffee cup. "I'm not worried about the future. God's got that. He was around at the beginning and he'll be around beyond the end. I'm more worried about the present. Emil says that somebody tried to break through our software firewalls again. People out there are interested in what we're doing, apparently." Despite the strong coffee and the warm setting, Todd shivered. He looked up. "Is Bradley Glenn meeting you guys at the airport?"

She shook her head. "We're meeting Dad there." She was glad Bradley Glenn wasn't going to be there. The drive would give her some time to steel herself. Why did people have to be so difficult? And what had her father possibly seen in this armchair paleontologist? "He sent us directions to the facility where the drawings are being preserved. We'll rent a car." She punched Todd in the shoulder. "Now who's being paternal?"

* * *

Shaving used to be such a simple thing, Xavier reflected. A quick swipe across the cheeks, a few scrapes down the neck—watch out for that Adam's apple—and he was done. But these days he had to watch for errant hairs in other places. There was the tip of his nose, where little hairs sprouted the same way they used to on his mother's schnoz. He particularly hated shaving his ear lobes, but he supposed it came with getting older. Small price to pay.

He couldn't see his reflection very well in the car window. The light was bad, but judging from recent visits to the washroom, that might have been a blessing. He could mentally check off the list: sunken cheeks, greenish circles under the bloodshot eyes, a general wiltedness to his demeanor.

The light turned green.

Xavier knew that he'd been overdoing it lately. Tonight's assignment: shut-eye. But if tonight was like the last few, he wouldn't sleep for long. The headaches ground away at him, a relentless pinch in the nape of his neck, a throbbing behind the eyes. It was a catchy drumbeat, but if he thought about it too long he became nauseated.

I should take a break from the Q-slips. From all the travel. After Princeton. Yes, after Princeton, a break. Just a few days or a week wouldn't hurt. He had to recover before they began their Thera explorations. He had to feel a little more energetic so he could rally the troops. That's what the leader was for, right? Rallying the troops?

Even if he'd had the time or the energy, he wouldn't have returned to ChronoCorp tonight. The place was beginning to give him the creeps. He kept seeing movement from the corner of his eye, phantoms flicking around doorjambs and crouching in the shadows. Sometimes he thought he heard a voice. Sometimes he thought it was his.

Morning would come too soon, and then he and Katya would be on the traveler's rollercoaster, rushing from point A to point B, straining to hear airport announcements over screaming children, weaving their way through various security checkpoints, tossing keys and cell phones into plastic bins as they tripped over their shoelaces, grabbing cups of bitter coffee and cinnamon rolls before boarding. Travel wasn't what it used to be. But that was tomorrow. Tomorrow, when all would be revealed in a simple set of drawings. A century of slumber in a crate. These pieces of painted cardboard, the leaves of scrawled paper, would free them to enter back into the quantum slips, or call into question every bit of work they had done for a decade.

Had Kat changed the course of time? Was the quantum structure of the universe more like a river or a pond? If a river, their Q-slipping changes to the timeline would fade away. But if the universe was pond-like? That increased the consequences, didn't it? Imbued every change with more import. He shuddered. Tomorrow would tell. He was just glad no one else had the technology. In the wrong hands, it could wreak havoc. And there had been rumors. Or were those his imagination? Drip, drip, ripple, ripple.

21

Blue Skies

Xavier parked in the long-term lot at Denver International. He'd insisted he could drive the route from Fort Collins blindfolded, but to Katya he seemed hyper-vigilant, constantly glancing in his blind spot as if to catch someone coming up from behind. In line for security, Katya kept a close eye on him. He seemed flighty, twitchy. His right foot tapped on the tiled floor while he wiggled his left hand in time to an inaudible tempo. To distract him, she launched into some small talk, and by the time they reached security, he seemed far more relaxed.

In yet another line, this one for coffee, Xavier rubbed the back of his neck. "I don't know, Kat. I have a feeling that either this drawing will have a profound effect, one that might have split off another universe that we won't ever experience, or that it will have no effect at all."

"Isn't the outcome the same?" Kat said. "I mean, if those are the two choices, it seems that anything we do will have the same result in our own universe."

"But you're thinking three-dimensionally again. We need to think in multiple dimensions. Is it our responsibility to make sure that we're not splitting off whole branches of reality in new directions?"

"Branches that we'll never see," Kat offered. "Besides, it may be that every decision we make divides into another universe, whether we're slipping through time or just living day to day. Lots of little buds and branches emerging from the quantum plain."

Xavier shook his head briskly. "Yes, we all know the drill. But we need to take a measured approach, and my vote is that if something at Princeton has changed drastically, we may need to cease our travels—as I suggested before."

© The Author(s), under exclusive license to Springer Nature Switzerland AG 2021
M. Carroll, *Plato's Labyrinth*, Science and Fiction,
https://doi.org/10.1007/978-3-030-91709-8_21

"We could change our protocols. I pushed the envelope in Central Park and admit that maybe it wasn't the best way of testing things out. But when we visit Thera, we can make sure we're unobserved, or that at least we make no changes to their world."

Xavier pulled on his chin, as if he could make his face longer and thinner than it already was. "Perhaps. Perhaps."

The flight was called, and they took their seats. As the aircraft climbed into the brilliant sky, Xavier pulled his thick glasses off and began wiping them on his shirt. "On the other hand, time is of the essence, so to speak."

Katya frowned. "I don't see the need to rush into anything."

"Au contraire, my dear."

"You're mixing your languages."

"If I'm not careful, I just might start mixing my metaphors. There is some rush."

"What's the hurry?"

Xavier's eyes darted back and forth for a moment, like a caged animal planning its escape. "Competition. There are others. Corporations. Organizations. Ones that are off our radar. Off the grid."

"Are they trying to shut us down?" Katya said with rising alarm.

"More like beat us to the punch. Time travel is the holy grail of physics. Slipping the quantum fantastic." He licked his lips. "Somebody does some grandstanding Q-slip before we begin exploration, before we publish, and I'm afraid Mila will pull the plug."

"But she's not after the glory. She's after the data. The history."

"It all seems that way," Xavier whispered, leaning toward her. His breath came in rasps. "But nobody's that pure. The joy of discovery is one thing, but fame leads to publicity and opportunity for someone like her."

"Was it Mila who told you about these supposed competitors?"

He bobbed his head to one side subtly. His left eye twitched. "Implied," he said. "Implied only. But I could read between the lines. I could."

The whites of his eyes looked like undercooked pork.

"Kat, who knows who's out there? Imagine our technology in the hands of some serial killer, or some dictatorship. Or terrorists!" Xavier clamped his hands to the sides of his head, as if to keep it from splitting open.

Katya felt the icewater drip along her spine. Her alarm wasn't only from the things he said, but from the way he said them. "Xavier, are you still getting the headaches?"

He reached over and patted the back of her hand, a quite unprofessional and un-Xavier-like thing to do. "They seem to come with the territory.

Headaches and flu symptoms. Insomnia. Q-slips are hard on the physical body and the mental state."

She met his eyes. "Just how many extra trips have you taken?"

He glanced down the aisle and silenced himself as a flight attendant passed by. As soon as the aisle cleared, he was back to whispering. "Just a few. Have to make sure of all the details. You know what they always say: 'safety third.'" He grinned, but a shadow fell across his face.

"Boss, I know something's up."

He looked out the window.

She pressed the point. "What's got you spooked?"

He turned and drilled her with his eyes. "I think I saw myself. After my last slip."

Katya let out a nervous laugh. "That's good. It proves you're not a vampire."

"Not in the mirror. In the corner of the room. He was standing behind me. *I* was standing behind me… Sounds crazy, right?"

He chuckled, but his laughter became a hacking cough.

The flight attendant was back. He said, "Sir, may I get you some water? Something hot to drink?"

Xavier waved him off. "I'm fine, thanks." The attendant walked on, but Xavier didn't take his eyes off him. From the side of his mouth, he said, "See how careful we must be?"

Katya stared at him. Xavier was right about one thing: he sounded crazy. Xavier trembled slightly, his entire body tense.

She opened her purse. "You want something for the pain?"

He looked quizzically at her. "I see the doubt in your eyes, my dear." His wide grin spread again. "Kat, as you know, just because you're paranoid doesn't mean they're not out to get you."

He closed his eyes, rocked his shoulders, and settled into his seat for a nap.

* * *

When Rex Berringer returned to his office, his sharp senses and training enabled him to spot the midnight blue truck half a block down on the far side of the street. The huge brushed steel vehicle was bolted together, with rows of sparkling rivets and silvery piping for good measure. Spikes protruded from its hubcaps. It was the kind of vehicle that didn't need turn signals; the traffic simply parted, like the Red Sea before Moses. He'd seen the truck before and knew exactly who it belonged to. He knew that in the back window was a mysterious sticker, a round decal with a warped cross. A sleek hood ornament of a coiled rattlesnake rose from the front. In the rear, the license plate declared, "Don't mess with Texas." Rex wasn't about to. He took these rendezvous very seriously. He just wished he knew when they were coming. Like the ghost of Christmas yet to come, the truck's driver was mercurial, showing up on his own schedule.

The truck-driving cowboy's bosses, whoever they were, seemed even more mysterious. They appeared to have an agenda with a serious timeline, but what it was, Rex had no clue. Although he'd not had direct dealings with them, they disturbed him even more than the Texan.

Dresden was waiting in the office, with his ashen face and his dark eyes. The man reminded Rex of something from the *Book of Revelation*, the pale rider of death and decay. Rex wasn't sure how he had gotten in, but Rex didn't ask. The man was standing, gazing out the window, studying Fort Collins like a microbiologist might examine a new slime mold.

"Mr. Dresden," Rex acknowledged evenly.

Dresden dragged his finger along the windowsill. "My friends call me Tex."

"Well, Tex, I'm—"

Dresden wheeled around, placed a firm hand on Rex's bicep and squeezed. The man had long, thin, bleached hands. They were impossibly cold and surprisingly strong. "I said, my *friends*. You may call me Mr. Dresden."

"As you wish, sir," Rex mumbled.

Dresden let go and turned back to the window, as if something out there was far more interesting than the business at hand. "I'm not sure you fully understand the gravity of your commission, both to your professional standing and your personal health."

"That sounds vaguely threatening."

"I'm sorry," Dresden said. "I didn't intend to be vague. My supervisors are paying you well. They are endowed with rather extensive financial resources, but they are also responsible with those resources. They don't like inefficiency." He crossed the room, sat down in Rex's chair, crossed his legs, and intertwined his long fingers atop his knee. "Now, what can you tell me?"

"I've visited the offices twice. Both times, the premises were unexpectedly closed."

"Unexpectedly? Don't they have regular hours?"

"It's not exactly a department store," Rex snapped. He quickly softened his tone. "On my first attempt, the principals had gone out. The second time, they had a short day. I do plan on making an appointment next time."

"That would be one recommended strategy," the pale man said. "Have you solved the issue of large files?"

"I've got a solid state flash drive I can smuggle in there that can handle the kind of files we want. This baby's got more than a terabyte. Very roomy. But I'll also access their internet directly, and ping the Imperium so we can access files clandestinely in the future." Rex liked to use words like "clandestinely." It meant secret. He had looked it up.

"Whatever you do, make sure you do it presently. The leadership is dying to know what's going on in there. I'm sure you are, too. Dying. To know."

"Dying? You do *dark* so well." Rex offered mirthlessly.

Dresden's lips pulled into a thin, colorless smile. "That's not up to me, of course. I'm just the messenger. Good day, Mr. Dolp."

Rex had never revealed his real name. Apparently, Dresden wanted him to know how much they knew about the private detective. It gave him a case of insecurity of the worst kind. But he found comfort in the fact that Dresden was not at the top of the food chain. Even the dark Texan had someone to answer to.

*   *   *

For all of his confident demeanor, Dresden had plenty of doubts. There were holes in the narrative he'd been given by the Ambassador. Big holes. Holes you could drive a Beluga Whale through. One concern was *why so much mystery surrounding the warehouse?* From the outside, it looked derelict, with dark, broken windows on the upper levels. Behind those windows stood walls painted black, a shallow façade hiding a fully renovated and well-lit interior. It was a clever disguise, enabling the building to blend in with its industrial surroundings. The exterior was just the beginning of the building's mysteries.

Its interior had bizarre furnishings. Dresden had asked about the Roman décor and its prominence in the Imperium's offices and facilities, but the Ambassador had been evasive, referring to artistic taste and a hobby of "the antiquities." But it went beyond the hobby level. The Ambassador often spoke in Latin. Yeah, a real history buff. But it struck Dresden as more practice for something later, as if the Ambassador was boning up on a foreign language before taking vacation abroad.

Dresden had little doubt that there was a connection between the Primus Imperium, ancient Rome, and ChronoCorp. But what was it? Why would ChronoCorp—a company that concerned itself with research—be so important to the Ambassador's goals? They were obviously not a bunch of history professors. The corporation was staffed with eggheads, mathematics types.

But the Imperium had their own experiments going on, and if it had to do with ancient Romans, Dresden suspected they had something to do with time travel. It was the stuff of science fiction, but what else could it be? Was it possible that ChronoCorp had made a technological breakthrough where the Ambassador's engineers had failed? How far had they gotten?

Dresden had a feeling he would be finding out soon. He had a feeling he might not like what he found.

22

Princeton II

The flight to the East Coast seemed longer than usual. Katya drifted off to sleep somewhere over Missouri. When the landing announcement woke her up, Xavier was staring at her.

"Hey, Sleeping Beauty, you had a Kat nap!"

Katya hoped she didn't look as strung out as Xavier did. He could probably have used a nap more than her. His hair stuck out in several directions, like a hedgehog, but not as cute. His eyes looked like the bulbous eyes of the Michelin Man. Their puffiness contrasted with the hollow quality of his cheeks.

Katya rubbed her eyes—they were probably bloodshot, too—and looked out the window. Somewhere below, her father was waiting. It had only been a couple weeks, but she couldn't wait to see him again. She wondered if he would still be caked in the powdered stone of Wyoming.

They disembarked into Newark Liberty International's Terminal B, a recent addition to the venerable airport. The place was polished glass and steel and tile. A tram took them to get their bags, and a short walk brought them to the rental area. ChronoCorp was paying the bill. At the far end of a long row of signs like *"We put you behind the wheel of our wheels"* and *"Drive in style without the price tag"* stood Ajit, sans any obvious grime.

He rushed to throw his arms around Katya, then extended his hand to Xavier. "Good to see you again, Dr. Stengel."

Xavier shook his hand brusquely. Instead of any pleasantries, he rasped, "I'll get us our car." His voice was a little too loud, and it wavered, as if he was trying to control it. He headed across the concourse for the counter.

Ajit looked at his daughter. He said under his breath, "What's wrong with him? Is he sick or something?"

© The Author(s), under exclusive license to Springer Nature Switzerland AG 2021
M. Carroll, *Plato's Labyrinth*, Science and Fiction,
https://doi.org/10.1007/978-3-030-91709-8_22

"He's had a lot of travel lately. You know how it is."

Ajit watched Xavier shuffling toward the rental row. "I wouldn't wish Highway Ninety-five on someone who's used to sleepy Fort Collins. He's having a hard time walking a straight line as it is. You sure it's a good idea to put him behind the wheel, dragging us along for the ride?"

Kat scampered after him. "Boss, maybe we should have two drivers, in case we need backup."

Xavier grunted. "Yeah, good idea. We'll put both names on it."

"Great. And why don't you let me do the driving first. I'm the one who got the nap."

He smiled faintly. "So true. I'd appreciate it."

They filled out the paperwork and found their compact car in the lot. Katya pulled her phone out for GPS, but Ajit said, "I know the way. I'll be your Google lady."

They lurched through clogged arteries of traffic near the airport, but once on the highway they got up to speed and enjoyed the view. A succession of communities passed by, each with its own unique character, modern, business, industrial, farmlands, and rural picturesque. The campus of Princeton finally appeared on their GPS, then out the windows. She pulled into an official space in a lot on Williams street, per Brad's instructions.

"Parking permit?" she asked Xavier. He was leaning against the window, mouth open, snoring.

Ajit reached up and tapped him on the shoulder. "Xavier, got that parking pass?"

"Where did I put that?" He fumbled through a file of papers and pulled out the sheet he had printed from the file Brad had emailed. They climbed from the cramped vehicle into the brisk New Jersey air and headed for the Seeley G. Mudd Manuscript Library, making their way past classical and modern buildings. The Student Center and Art buildings presented a retro-contemporary feel, but beyond them rose brick and rough-cut stone edifices, a high gothic chapel, and formidable dormitories.

"Ivy league is right," Kat said, looking at the greenery growing on the walls.

As they reached the entrance to the Library, Ajit held the door open. Kat stepped through and looked…up. Light beamed in from clerestory windows, and a series of glistening bookshelves held multicolored tomes within polished wood and snow-white dividers.

"Lovely, isn't it?" Ajit said, gazing up. But Kat had dropped her view to the lobby, where people milled around or sat at study carrels with headsets. On the far side of the atrium stood a man in a collared shirt and jeans. He was thin and wiry, in the same manner as Xavier, but a bit more athletic from the looks of how he stood. From her father's descriptions, she knew it was Bradley Glenn. She tugged at Ajit's sleeve.

"That him?"

Ajit laughed and called out. "Brad!"

As the men shook hands, Kat could see that he wasn't much to look at. He was, in a word, gangly. Not that it mattered to the task at hand, but her father had built him up in her mind to the point that she was expecting him to have morphed over the years into someone along the lines of Indiana Jones.

Ajit introduced everyone. Katya studied him clandestinely—dubiously— as he and her father caught up on life. He did have a cute quality to him, and the kind of fire in his eyes that spoke of a lot going on in that braincase. Maybe they could get along after all.

"So," Brad said, "Doctor Stengel. Dr. Joshi and Dr. Joshi. So many PhD's to welcome. Glad you're all here. Shall we take the scenic route up two flights, or take the elevator?"

"The scenic route, always," Kat said.

Xavier blinked toward the elevator. "Think I'll use the local technology."

Ajit said, "I'll go with Xavier, and see you two up there." *You little lovebirds,* she could hear in his tone. It was annoying.

Ajit and Xavier set off toward the elevator. Brad shrugged, gestured toward the staircase, and said, "Shall we?"

He had that look that she'd seen before, a look that said *this is not that scrawny kid in a pretty dress that I saw all those years ago.* She knew her dedicated jogging and kickboxing had sculpted her body in ways that men found attractive, and she wasn't self-conscious about it. She had even used it to her advantage, though rarely. Still, his expression amused her to see, especially on the face of a young colleague of her father's.

As the two ascended the stairs, Katya noticed a tablet in his hand. It wasn't an electronic one, but rather an old-school paper sketchpad with a pencil clipped to one side. She wondered what was inside its pages. Brad was in tour guide mode, pointing out various displays and architectural features. There was an edge to his narrative, a certain tentativeness. He seemed to be intimidated by her. She felt sorry for him.

"I bet you've taken these steps a few times. You seem used to them," she said.

"I visit a lot for research, but where I work they have me in the basement. Those stairs are easier. If you're really in a hurry, it's just 'tuck and roll.' But I cycle to work and I ride in the country on weekends. I was in a race last month."

"One of those three legged deals with a potato sack?"

"Something like that."

"How'd you do?"

"We won't speak of it. I do it to retain my Greek-god-like physique." He said it like he didn't believe it.

Talking about his physique was the last thing she wanted to do. She pointed to the sketchpad. "Do you draw?"

"I find that it helps me see. We rush through life without really watching. We miss a lot. This little pad helps me slow down and take it all in."

"May I peek?"

He offered her the pad and leaned back against the bannister, trying to look disinterested. But she could see his furtive glances.

Katya leafed through sketches of the interior of the building, and then a few studies of plants and insects. The most recent page displayed a Latina woman in profile. He was good.

"Impressive," she said, handing it back.

"Thanks." He stepped onto the final flight of stairs.

"Is that someone special? The woman on the last page?"

He paused, his foot on the first step, and opened the pad again. "My coworker at the museum. Delaney Delgado. The way she's stationed at her desk, it's easy to draw her profile. We hired her right after she finished an internship. Very sharp. And easy to work with."

Katya looked over his shoulder at the little portrait. She was tired from travel, and the calories from the small granola bar served on the flight were long gone. Maybe those things contributed to her dark mood, but Brad's tone seemed perfunctory, judgmental. "Have there been people who were *not* easy to work with?"

"Oh, we've had some loonies. This one girl from the Bronx was certifiable. She fabricated most of her resume and actually stole some funds. And then—"

His voice trailed off as he locked eyes with Katya. He had the same look she remembered when he spilled punch all over her all those years ago. Brad said, "Well, that was an unkind way for me to put it. I suppose all of us…"

Before he could finish his sentence, they arrived at the third floor, where Ajit and Xavier were waiting. Just like their last phone call, they were cut off by circumstances. But it didn't matter, did it? If some kind of relationship was meant to be, it would take a severe transformation on the part of Bradley Glenn. Katya was far less concerned than her father apparently was.

Professor Chad Markham stood behind Ajit and Xavier. He must have crossed campus specially to welcome the scientific entourage. Ajit made introductions again. Markham turned to Brad. "Well, it's worse."

Brad grimaced, recovered, and said, "Can we see your historic drawings?"

"Absolutely," Markham said. He knew Ajit and the AMNS group were the most invested in the images themselves, but Kat realized that both Markham and Brad had little idea where Xavier's and her interests lay.

Markham led them down a corridor to a darkened chamber. The paintings had been moved to a special climate-controlled room. Cool air rushed into grates in the floor. "We keep the moisture down and the temperature steady here. We also ask everyone to wear gloves. Some of the surfaces are very delicate." He doled out protective gloves like cards at a poker game. Then he gestured to a chair and said to Brad, "Please, take a seat." He handed him a magnifying loupe.

Markham spread a series of matted works across the table in front of the group. In moments, the art of Waterhouse Hawkins transported them to another age, a deeper time. Here, a lumbering Megalosaurus opened toothy jaws at a bat-winged flying reptile. There, a sunset blanched the scaly skin of a flippered plesiosaur splashing in a primordial swamp. A long-necked Apatosaurus stretched her neck to peek through the treetops.

Kat peered between Ajit and Xavier. Brad kept laughing and pointing and blinking. "Amazing," he murmured. "Just amazing. I've been a fan of this guy's art for a very long time. And now we find originals locked in a crate. They're thinly sliced history. That's what these are."

Kat gestured toward the line of drawings spread on the table. "Seems like some of these are way too small for their mats."

Markham said, "These are not for exhibition, so we treat them differently. When we mat for storage, we make the mats a uniform size, so they can stack together into a fitted box. Theoretically, these works could be stacked on each other in a storage box for decades. If we sized each mat to its drawing, then a pile of them would mean stress and deformation of the art works as they rest on each other. That's why some of the smaller renderings end up swimming a bit in their archival mats. It's all about protection. We mount each one in a hinged museum board frame for easy access." He lifted one of the facing mats to expose the drawing and then carefully swung it back into place.

Xavier took off his glasses and leaned over one of the renderings. "Beautiful." He straightened. "But these are standard fare for the time. Precious, but consistent with what we've seen of Hawkins' work. Where's the other one?"

"Yes, the one you came all this way to see." Markham said. "It's in what you might call our Intensive Care Unit."

A woman entered the room wearing a hood, mask and gloves. "Guests!" she said, surprised.

Markham said to the group, "This is Dr. Evie Long, one of our top curators, dressed for Halloween."

She pulled her mask down. "Sorry about that. I've been in the archives down below. That's where we keep the really old stuff. Dusty, too." She tapped her mask.

"Can you give us an update on the Hawkins piece?" Markham asked.

"Biggest mystery I've ever seen. It's as if someone has misted it with acid that's slowly eating it away. Have you ever had a really old bottle of wine? As soon as it's uncorked it instantly begins to turn to vinegar. You've only got a few precious minutes to taste it." She hooked a thumb toward the doorway. "As soon as this thing hit the air, it seems to have begun to disintegrate. We saw subtle pits at first, and took several steps to preserve the material and dyes, but the surface continued to degrade. Come, see for yourselves."

Katya caught Xavier's eye. *It wasn't the air.*

Dr. Long opened double doors at the back of the room. "No need for protection here. Everything's sealed under glass."

The room had no apparent interior lighting. Shelves lined the walls, and several tables slumbered in the dark. As their eyes adjusted to the dim room, Dr. Long said, "Do you see anything?"

Katya let her eyes relax. She scanned the darkness. Toward the middle of the chamber, a blue glow emanated from a table. It was the same blue hue she had witnessed in her Q-slips. The room blew through seasons, summer to winter in a flash. "Blue," she muttered. Xavier nodded. He'd been there, too.

Long said, "Oh, yes. But it's fading, just like the structure of the paper itself." She flipped a switch. The room remained dark, but lights illuminated the table in the middle. "Here 'tis."

They huddled around the glowing counter, a neon casket holding a dying painting of a dinosaur. And it was everything they had been told. Both Ajit and Brad took in a breath as the image became visible. Ajit pointed to the erect tail. "Mm," Bradley nodded.

Then it was Brad's turn to point, to the creature's thumb spikes. He turned to the group and said, "In Hawkins' time these were thought to be a horn, like that of a rhino. How'd he ever guess that they went there, on the manus?"

Kat felt a current of chilled air, but it had nothing to do with the temperature of the room. Whatever changes she had put into play had made it at least this far. She had rocked the great hourglass of the universe backward, and the sands of the future had mixed with the sands of the past.

Markham's soft voice betrayed a sense of awe. "He put the spikes in the right places."

Brad held up the loupe and hunched over the table. He brought the eyepiece nearly to the surface of the glass and scanned across the length of the drawing. Then he took another look at the brush strokes. He kept staring through the lens, but asked Long, "Was there any watermark on the paper?"

"Yes," she said. "It's got an N. Pannekoek watermark, with a rampant lion."

"Crown above?" Brad asked. *How much did this paleontologist know about paper?* Perhaps it was simply the fallout from studying an artist like Hawkins.

Long nodded. "And it is 'wove' paper, not antique laid, so that fits the chronology as well."

He stood, jammed a fist into the small of his back, and looked at the group. "Well, it certainly looks like the genuine article. Drawing materials and paper are all the right time period." Brad looked at the ceiling. "It's like someone told him." Even in the dim light, Kat could spot the realization on Brad's face, followed by a flicker of shock. "That's exactly what it's like, isn't it?" The question wasn't rhetorical, and it wasn't directed at the general crowd. He was asking Katya.

Kat looked down at her shoes, wondering how much to say. Why couldn't he leave things alone?

Xavier derailed the question by asking another. "So, Dr. Long, what can be done about this poor dissolving drawing? Can it be saved?"

Was it Kat's imagination, or were the pits even larger than when last she looked? The color seemed to be fading as well, the blues graying, the reds bleaching to an anemic orange.

The curator turned her palms up. "I've done everything I know to stabilize it. I've consulted people at the MOMA in New York, and I called a colleague in Germany who's very good at these things. They're all stumped. They advised taking a suite of photos as soon as possible."

"And did you?" Kat asked.

"Something went wrong with our photographers' equipment. They're going to have to do another set."

Kat looked at Xavier. His eyes were closed, but he was nodding that he understood, or perhaps that he expected as much. She scanned across the room. Bradley held her gaze for a moment. He was trying to work something out.

"Well," Chad Markham said firmly, "good luck on that. Does anyone have any questions?" The group stood in stunned silence. "If not, we'll leave you to your duties. Thanks so much, Dr. Long."

Everyone else mumbled their thank-you's. They filed out of the room. Brad cornered Kat as she passed by and whispered, "Just what are you guys up to?" He spat the words as if they had fury behind them, but he was keeping things quiet for now.

She stiffened. "Can we talk about this later?"

"Honesty is the best—" He stopped himself, apparently realizing that he had blurted out the words. He lowered his voice for the rest, "—the best policy. I prefer to keep everything above board, in my scholarship and in life."

Katya bristled. She hated feeling defensive, because at those moments she usually said stupid things. "Gee, it must be nice to be so above board that you're so above it all."

She stepped quickly down the hall and joined the others, loitering until Xavier caught up. Then she stepped over to him and put a hand on his elbow, glancing around to make sure no one was listening.

"Boss, do you think we'll need to be working with the paleontologists any more on this?"

"It's certainly possible, depending on what happens to that poor little drawing."

"I'm sure we can find someone who is as much of an expert as Dr. Glenn to help us out. Someone more local. Maybe from the Denver museum? Or even Montana—Museum of the Rockies?"

"Actually, I was thinking of inviting Bradley out to see the facility."

"What? Why?" Katya hissed through clenched teeth.

"Bradley Glenn is just smart enough to know something is up. He might be smart enough to figure out more than an amorphous hunch. Something has to be done about Bradley Glenn. He needs a keeper. We need to keep him under control."

"The old 'friends close and enemies closer'? I'm not so crazy about asking him into our inner sanctum at this point. It's premature, if not rash."

"Believe me, I'm not that crazy about the idea either. But better to have him on our side than spreading rumors, true or otherwise. Besides, since the Hawkins drawings are a sort of control for us, a yardstick of the changes we've made, who better to help us evaluate that than Bradley Glenn?"

This was not what Katya had in mind at all. Having Bradley Glenn traipsing around ChronoCorp would be annoying, to say the least.

Xavier rubbed his eyes. "But you do it. You know him better than I do. Just do it when nobody else is watching. I don't want a bunch of Boy Scout troops and tourists baying at our door. At least not until we've had teams on the ground at our target and we've published something."

She sighed heavily. "Understood. We're not public yet."

"Good girl."

Usually, that phrase made her recoil, but she was distracted by the business at hand: how to get Bradley alone? She might have to do it remotely. Email was dangerous. A text? That still could be intercepted. She needed to do it in person, and before he ran off to complain to someone. He was becoming such an irritation already.

Katya made her way back through the parade of people and tapped Brad's shoulder. "Can I ask you a technical question?" Markham glanced back at the two of them. "In private?" she added.

Brad scowled at her. "Sure."

The two stepped back into the other room. Katya tried to keep the acid from her voice. "Look, I'm sorry I snapped. It's been a long day already."

"Travel has that effect on me, too," he said.

She softened her tone. "Okay, Einstein, would you like to come see what we do? Just you?"

Slowly, suspicion turned to interest. "I've never been to Fort Collins. Sounds like a nice place."

"That it is. We've got to get a few things done back at the lab before you come out."

"Lab?"

"We'll be in touch. But it's important, at this point, that you keep this all under your hat."

"I've got a Yankees cap that will do just fine."

"You can show me some more of your renderings," Kat suggested as they caught up to the rest of the gang. Xavier was watching her with something like amusement.

For now, Kat had put off the potential problem. But it wouldn't last. ChronoCorp would have to make good on her promise, sooner than later.

23

Phantom Visit

On the drive back, Ajit slept in the back seat, the boss rode shotgun, and Kat was at the helm again. Xavier rolled his eyes at her. "Want to see my etchings?" he intoned. "Please tell me this is nothing more than professional."

She glanced in the rearview mirror. Her father was still asleep. "Nothing more than professional. Besides, the way he talks about his colleagues drives me crazy."

"And why would that be? I mean, why do you care?"

It was a good question.

"The only reason you would have any opinion about Bradley Glenn's behavior at all is if you were interested in him."

In a teasingly paranoid voice, Katya said, "Who sent you here? You sound like my father."

Xavier leaned back, but he didn't close his eyes. The sight of the Hawkins drawings seemed to have energized him. He began to whistle what sounded like some tune from a Broadway musical. Finally, he paused his little concerto and said, "We're playing with causality here. Like Mila said, we get used to thinking of the events in our lives—little ones and major ones—as having an order. One after another. But..." He pointed upward. "Did you read that stuff by Romero's group?"

She sensed a nerd moment coming up. Sometimes she felt as though Xavier was testing her, seeing if she was keeping up with the scholarly articles. And while her life seemed to be filled with distractions these days, she was familiar with this paper. Thank God. Test passed.

"The Brisbane study? I just scanned it. What was the gist?"

© The Author(s), under exclusive license to Springer Nature Switzerland AG 2021
M. Carroll, *Plato's Labyrinth*, Science and Fiction,
https://doi.org/10.1007/978-3-030-91709-8_23

"They used polarized light for a double operation. Either the light inter-acted with the first and then second operations, or it did them in the opposite order. But instead of experiencing things in a linear way, the light could actu-ally carry out the two different outcomes simultaneously."

"Superposition, yes."

He leaned forward. "And maybe that's what's happening here. With the changed drawing and the originals."

Kat didn't try to hide her bewilderment.

Xavier became more animated. "Kat, look: Hawkins has now done a draw-ing, directly because of your visit, that depicts an Iguanodon in a pose embraced by modern paleontologists."

"Modern paleontologists don't walk like that," she objected.

"You know what I mean. So we have a causal effect that has issued from your Q-slip. Perhaps we are now living in a reality split slightly from the one we're used to."

"Maybe," she said, scratching her cheek and frowning. "Or maybe this is one of those cases you talked about where its imprint is too faint to last."

"Yes, hence the dissolving paper. Either way, we're on to something new. Something we must pursue academically and in the lab. More testing."

"More Q-slips," she added.

It seemed so ironic. Xavier was counting on time being unchangeable so that he could safely carry out further research. She was banking on making her own world—everyone's world—into a new universe, a place where today's technology fed the past, and that altered past in turn enhanced the present. Katya remembered her mother's reddish-brown hair, a trait she had passed down to Katya's own coiffure. She could go down that trail, remembering many things about her mother: her fragrance, her softness, her stubborn search for a balanced world. There was no way around it: the universe would be a better place with her than without her.

*   *   *

Xavier sat at a table in the employee lounge, staring at the tepid cup of cof-fee in front of him. These days, coffee only made the shaking worse, and over-the-counter meds hadn't touched his headaches in days. The trip to Princeton hadn't helped.

He had been thinking about their Minoan scholarship, and he didn't like what he was thinking. They had no technology that would actually document the history they were seeing. They could be good observers, yes, but how would their results be perceived in academic circles? What were they going to do, wave their hands around? Circulate "artist's conception" press releases? It wouldn't do. The techs had another system to try—what they were calling a "bioimager" made of cloned skin material, sinew, and living bone tissue. It had the appearance of a deer's foot, with lobes on the furry sides and a row of orifices along the top ridge. At one end, an opening accessed a lens, a mass of eye tissue complete with internal retina and optic nerve. The thing looked like something out of H.P. Lovecraft. Its electrical impulses, stored in a small mass built of neural tissue at the back, just might record the ancient world around them. Todd called it a "platypus polaroid."

It didn't help Xavier's mood that Bradley Glenn was coming out for his tour. Xavier had come to realize that he didn't trust the good Dr. Glenn to keep this kind of secret. He didn't trust anyone.

It was time for another trip, and that trip would be tonight. Everybody had gone home hours ago. It was Friday, after all. One late-night sojourn to a known quantity. A return to nineteenth-century Central Park should do the job; Katya had already led the way. One more solo voyage just to check it out, smooth the wrinkles and test the bioimager. Then, a tandem with a second person. If dual time traveling worked as it should, ChronoCorp would be ready for Thera.

Tonight's Q-slip would feature the first tandem. He didn't need a building full of people to monitor a simple trip like this, and he didn't want to wait. He could feel the competitors breathing down his neck. He would go with an inert package in pod #2. He had it all worked out. If all went well, he'd return with that goopy, pulsating organic camera and actual images of historic New York.

"You really shouldn't do that," came a quiet voice from behind him. It had a strange, echoing quality to it, as if coming from the far side of an empty sports stadium.

He didn't want to turn around to look. He knew the voice. It didn't belong here. Eyes focused on the counter in front of him, he said, "Do what?"

"Travel. Exploration is for teams, not for solo. Look, buddy. I came to warn you. It will make you crazy. Don't go. You're playing with quantum fire. *We* are." The voice was thin now, nearly too faint to hear. "Watch out for the Primus Imperium. They're the ones you need to worry about. And above all, don't go back."

Xavier swung around. The room appeared to be empty, but he knew it wasn't. He felt a presence—something in the room that didn't belong. He caught a glimpse of something in the mirror above the sink. It was the reflection of a tall thin man with thick glasses, not sitting at the counter with a cup of cold coffee, but walking away into a waning blue light.

He was living in a modern horror story. He had to get out of the room and finish what he'd started.

Xavier sprang into action, sprinting down the hallway to the Transport Arena. The monitors were already up. He hit several keys on one console and placed his package into Pod #2. Then he sequenced the other pod, grabbed two control units hanging from the wall, and climbed into Pod #1. He glanced at the shovel on the wall. *Sometimes it just piles up dark and deep.*

The first time he had embarked upon a Q-slip, the trip itself was disappointing. He expected some kind of rushing tunnels or flashes of light swirling around him. Instead, as Todd had put it, the slip was a bit like falling off a chair. One moment you were looking at one view, and the next you were on your ass, disoriented and looking at something else. For Xavier, those arrivals were becoming more and more disorienting.

At first, Xavier thought something had gone wrong with this most recent slip. He looked up and could see only darkness. Had the power failed in the lab? But his eyes began to adjust as the blue light drained away, and he could see stars beyond a silhouetted tree spreading above him. He popped the lid. So far, so good.

As soon as he cracked the seal, new smells assaulted him. Aromas of smoke, horse dung, and urine stung his nostrils. The air was heavy, mephitic. He cringed.

Xavier stepped out and spotted the faintly glowing Pod #2. He stepped over to it, unsealed the hatch, checked to see that the package was intact, grabbed it, and then closed the lid again. Now he looked around. It was unfortunate that he had arrived at night. He would have traded the privacy and seclusion for a bright afternoon, because he needed to test the bioimager. He glanced from side to side. He had ended up in a small stand of trees, with his feet firmly planted in dead leaves and twigs. But there was light. It filtered through the leaves farther on. He snuck toward it. Between the trunks up ahead, he saw a paved footpath. Further down, the path was lit. He peeked out and spotted a gas lamp in the center of a pool of light. He snapped a photo. The imager made a squishing sound. It was slightly gooey. He ventured further down the path. A driver sat in his carriage, his back to Xavier, just beyond the gas streetlamp. Xavier snapped several other images. Then he turned the organic marvel to the sky. With the slide of a valve-like switch, he changed the exposure to image the stars. The images would provide the techs with a host of data about time of arrival and even season of year.

He scampered back into the shelter of the grove, hit the button on the second talisman, and tossed it and the bioimager into the second pod. He climbed into his pod, folded his long legs up and pressed the button on his talisman. He found himself back in the darkened lab. Lifting the hatch open, he swung his legs out. The floor tilted to the left, and he saw spots in the dark room. Then, oblivion.

*   *   *

Xavier awoke to a lightless Transport Arena. As soon as he got his bearings, he lunged for the light switch by the doorway. Light flooded the room. He rushed to Pod #2. His package had made it safely back with him. He found the disgusting organic imager on the pod's padded bed where he had dropped it. *Platypus polaroid.*

The room was still tilting. His vertigo wouldn't go away. He felt nauseated, and his headache was unbearable. Stumbling to the lab, he plugged the bioimager into the system the way the techs had shown him. He hobbled to the lounge, popped two tablets and swilled water from the sink's faucet. His coffee cup was still on the counter, still cold. He couldn't drive.

He opened the Lyft app to book a ride home. The light of dawn trickled through the widows down the corridor. He had to get out before anybody came in and discovered him. He trusted Todd and Katya, of course. But he did not want anyone else to know. The walls had ears. For some time, he had wondered if their building was bugged, and if they were being surveilled. These were the things of spy novels and thriller movies; he knew Katya would dismiss such notions, so he'd decided to keep them to himself until he had evidence. Whoever he or she was—or *they* were—they would undoubtedly show themselves. One day. His team must complete their work before that happened.

While he awaited his ride, he thought back to his spectral encounter in the employee lounge. He texted Todd.

TODD: CAN YOU PLEASE DO AN INTERNET SEARCH FOR THE IMPERIALS OR THE IMPERIUM, SOMETHING LIKE THAT? MAY BE IMPORTANT.

He took the elevator to the ground floor, exited to the front sidewalk, and leaned over a raised flowerbed to lose his lunch. He closed his eyes and waited for his ride.

24

Clandestine Chatter

Emil usually arrived at ChronoCorp in the late morning, but today, he got an early start. He prided himself in his work and seldom found the need for human, canine, or feline companionship, although he did have two parrotfish. Radium and Tritium were so simple. Toss a few flakes in, keep the green fuzz scraped off the glass, and they were happy campers. The human psyche rambled among all those complex personality types: sanguine versus melancholy, lion against beaver. But computers—his endeavors—were the best. They relied on the black-and-white world of zeros and ones.

As he surveyed the lounge, he wondered what kind of personality type left the room in such disarray. Half-full coffee cups perched on the counter above a large, dried puddle on the floor. Footprints led from the stain to parts unknown. Food-encrusted silverware filled the sink, and a trash bin overflowed with gooey paper goods. It was all probably the work of one of those sanguine types. ChronoCorp had several. He nuked some coffee and headed for the lab.

On his workbench, Emil found something even more surprising: Someone had plugged his bioimager into the system. He had only gone over the details with Jinny, his most trusted tech, and Xavier. The bulky bootprints on the floor ruled out Jinny.

The imager had been set to download, but had not yet been triggered. Perhaps Xavier was just trying it out, seeing if he could remember how it fit into the hybrid interface. There was only one way to know.

Emil enabled the system and tapped a lobe on the imager. It felt a little like tepid, raw beef. His finger stuck to it slightly. A patch of fur on one side rose briefly, like the hackles of a dog. The bioimager shifted a little, sliding on its

© The Author(s), under exclusive license to Springer Nature Switzerland AG 2021
M. Carroll, *Plato's Labyrinth*, Science and Fiction,
https://doi.org/10.1007/978-3-030-91709-8_24

own patch of slime. Emil wiped his hand on his sleeve and sat back. He had reached this point many times before, downloading various devices from other Q-slips, only to get static or photos that faded away before his eyes. But the receiving computer beeped obediently as the first image materialized. At first, he thought it was a blank, or impossibly underexposed, but then he saw the faint squiggles in a familiar pattern. He knew instantly what they were: the smeared star trails of Orion. Someone had taken a picture of the night sky, just as he and several other techs had suggested. The second image scanned across the screen: cobblestones beneath a streetlight. But this lamp was gas-powered. Yet a third came in—unprecedented. It revealed an old carriage, the reins leading up to a hunched driver. A forth shot rounded out the collection, a clear portrait of two time pods in a grove of trees. Two pods. Xavier must have taken the extra one along, and these photos, unlike all their predecessors, weren't going away.

* * *

Kat took the elevator to the offices of ChronoCorp, whistling tunelessly. She never was musical, the only one in her family cheated of a singing voice or a talent for strings or brass. On the other hand, as her father was fond of pointing out, she was the only one in the family blessed with a mathematical mind. Her mom had forbidden her dad from balancing the checkbook back when Kat was in middle school and had offered Kat the job instead. She remembered the incident well; it was a highlight of family history.

Emil was waiting for her at the front desk. "Katya, can you come back to the lab?"

"Lead on," she said, following.

Walking briskly down the corridor, Emil stage-whispered over his shoulder. "I think we're about to do something really inadvisable."

"Inadvisable sounds fun. Who's doing the advising and who's ignoring it?"

"I'm ignoring Xavier's suggestion that I should leave him alone, and Xavier is ignoring my twenty-six years of life experience and wisdom." Emil paused and turned to her. "He came in here this morning looking like roadkill reheated a few times."

They turned into the lab. Xavier was there, slumped by the bioimager. "Well, hello you two," he said hoarsely. Katya noticed his untamed whiskers, the red rims around his eyes. His entire body looked wilted. "Looks to me like we've discovered a winning combination in our bioimager. Look at these, Kat! Seem familiar?" He jabbed a finger toward the screen. On the monitor, an image of cobblestones in 19th-century Central Park glowed steadily.

Katya leaned closer. "Looks really good. Are there others?"

Emil nodded.

"And are they all persistent, or are any disintegrating?"

Todd stepped through the door. "All there. No disappearing acts."

The laboratory had no windows, but Xavier glanced toward the door. Reaching across the desk, he flicked a switch and the monitor died.

"Hey, I wanted to see the others!" Kat snapped.

"You've been there, you've seen it before." He looked toward the door again and licked his dry lips.

Kat searched Emil's expression. He was shaking his head almost imperceptibly, but enough to communicate the unease he felt toward the boss. Xavier's hypervigilance was certainly warranted for the work they were doing, but their security was bolstered by the fact that they stored no data on the cloud. Instead, they had an off-site server to back up the data, and even the information housed at the lab was heavily encrypted. So why did Xavier seem so wound up?

Xavier lowered his voice. "Todd, did you have the chance to look into that thing?"

"Yeah, I'll shoot you my notes. Not much there. Primus Imperium deals in some low-end industrial research. Some pharmaceuticals. I did find one interesting thing: their leadership has been taking an interest in ancient history, just like we have. Maybe it goes with being an eccentric in the tech field. You guys might have a lot in common."

"That's what worries me," Xavier frowned. Katya caught Todd and Emil exchanging glances.

"So," Xavier nearly hollered. "Listen up, people. This paves the way for Todd to document his Q-slip to France. Knights and ladies. Castles and cathedrals. And after that, we go to Atlantis. Kat, you must put Bradley Glenn's tour off a little longer, just until we prove we can make the Thera connection safely."

Katya looked toward some invisible horizon. "Bradley Glenn. What's with that guy? He's like a bloodhound."

Xavier said, "Blame your own father for siccing him on us. But keep in mind that he's our control; he'll be able to tell us about any changes in the Hawkins part of the timeline."

Todd held up his hand. "France is fine, and I'm excited to go. But Thera still seems premature. No one's ever gone back this far, and we still haven't tried a tandem voyage. We're talking thirty-six hundred years, as opposed to your century back to New York. We may get quantum drift, or some kind of timeline warp. Right?" He looked to Emil for backup.

Emil shrugged. "Actually, everything's ready from our standpoint. The hardware's good; software's good. And now the wetware is good." He gestured toward the pulsating bioimager.

Todd said, "Thanks a lot, Emil. I've got your back, too, buddy. Just wait and see."

Emil leaned toward Xavier. "Make no mistake: I'd still like a couple more short rehearsals before we do a deep-time dive. I'm concerned about tandems. We really don't know if both pods will arrive concurrently. Why not send a couple of chimps or something? It worked for the Mercury program."

Xavier rubbed his hands together. "The pods did just fine on my trip to Central Park. Not to worry. I suppose there's no time like the present!"

Katya said, "I wouldn't climb into that thing just yet. Not to Thera."

Xavier stood. "You're right. You won't. It's going to be Todd and me. With Todd's architectural minor he can craft our first reports using words like 'cornice' and 'portico'."

"Yes," Katya reconsidered. "I suppose if you wrote it, we'd be stuck with words like 'nurnies' and 'thingies'."

"Exactly," Xavier said. He reached over and put a hand on her shoulder. "Kat, I trust you. I want you here in case the quantum shite hits the fan. If we don't make it back, you're our spokesperson. You're better at it than anybody else here. If we get back in one piece with intact images, it will be your turn for some real, meaty research. And after that, we publish in the very best journals."

Kat looked longingly down the corridor. The Transport Arena waited, but it wasn't waiting for her. Not yet. How she wanted to see those things Todd would see in France, and the Aegean waters lapping the shores of ancient Thera. She hoped her time would come, but it would only if Xavier and Todd made it back in their right minds with good data. A lot to hope for. Emil was right. Even if France went well, the Thera voyage was premature. She prayed it wouldn't end in disaster.

* * *

Katya's phone rang as she was tossing another "Microwave Masterpiece Insti-Chow" into her kitchen trash can. "H'lo?"

"Mila van Dijk for Dr. Joshi," a business-like man said.

"Oh, yes," Katya said, taken aback. "Sure. I'll take it."

"I'm so glad you picked up, my dear." Van Dijk sounded genuinely happy to connect with Katya. "Are you in a private place?"

Kat glanced toward her cat, Wellington, who watched her from the stuffed chair in the living room. "The environment is secure except for one feline."

"Ah, yes, I should have known you were a cat person. Yet another trait we share. Now, Katya—may I call you Katya?—I have something rather uncomfortable to broach with you, a favor to ask, woman to woman. Are you game?"

Katya was genuinely puzzled. This didn't sound like a business proposition. What could ChronoCorp's benefactor want? "Sure, I'm game," she said cautiously.

"I don't have to tell you how special a person Xavier Stengel is. The man is brilliant, not only as a physicist but as a human being. I find him to be intellectually challenging."

"He is that," Katya agreed.

"Do you know if he has any…entanglements? Anyone special in his personal life?"

Katya felt as if she was aboard a tall ship, just broadsided by an invisible cannon ball. She hadn't seen it coming. Was Mila's sponsorship of their research a product of infatuation?

"Xavier…er…yes. I mean, yes, he is…unencumbered. He had a romantic relationship a couple years ago that went south, but these days he fits the stereotype of a man married to his work."

"I see." Mila became quiet for a few moments. "Living in this mansion, I lead a fairly solitary existence. What would you think of arranging for Xavier and me to meet in his office now and then? I had hoped by now to make some social inroads with the man, but as you say, he seems married to his work. It would be easy for you. Tell him I'm heavily engaged in this project, both financially and emotionally, and I have had some degree of frustration getting a feel for the progress of the group."

"I do suspect he hasn't told you of all our Q-slips," Kat said carefully. "Some were simply technical forays. He probably didn't want to—" she searched for the right phrase, "burden you with information that he considered non-critical."

"I appreciate your loyalty, but please, Katya, burden me. I have more than a passing interest in Xavier."

Ahh, Katya thought, Xavier's love life might just be picking up. The idea of Mila van Dijk and Xavier amused her. What could it hurt? "You have a deal, Ms. van Dijk."

"Please, call me Mila."

25

Todd in Wonderland

The first thing Todd noticed was the light. In Colorado, it was nearly noon, but here the sun's rays had just begun to kiss the countryside with purple and orange. He sat up and waited for the vertigo to subside, rubbing his temples. The songs of meadowlarks mingled with distant church bells. There was a bite to the early morning air, and a taste to it: smoke, wildflowers, tilled soil. He had made it—he had safely arrived in the medieval period of the Kingdom of France.

Todd's pod had come to rest in a meadow, just at the edge of a grove of beech and chestnut trees so common to the Aquitaine basin. Confirming their aim point, the river Garonne snaked its way across the landscape. He was in just the right spot, and that was worth a lot. Beyond the river, behind a bank of distant trees, lay the smallish medieval town of Toulouse, its columns of smoke marking the fires of people preparing for their day. In a few centuries the town would grow to be the fourth largest city in France, but in this time, it stood as a sleepy little village.

He wished he were closer. He longed to see the Pont Neuf Bridge freshly built, the Toulouse cathedral, the convent of the Jacobins, Saint-Sernin Basilica. Todd reflected upon how many structures dedicated to God had withstood the test of time. People had built them as monuments to something bigger than themselves, something to outlive their generation. Many of the religious structures standing today would stand in his time as well. Saint Thomas Aquinas had died not far from here in both space and time.

Another bell pealed, this one more distant and deeper. He stepped from his pod and pulled out the slimy bulk of the bioimager. Would it work a second time? He shot the landscape, but aside from a low stone wall, there was

© The Author(s), under exclusive license to Springer Nature Switzerland AG 2021
M. Carroll, *Plato's Labyrinth*, Science and Fiction,
https://doi.org/10.1007/978-3-030-91709-8_25

nothing of note to photograph. Gingerly, Todd set off toward the river. It wasn't in Xavier's plan. The "architecture five" outline called for a quick look around, some test shots, and an immediate return. But Todd was an explorer at heart. He wanted a better look at his beloved medieval France.

A dirt road split the meadow. He paralleled it, staying in the tall grass in case he had to hide. Although he had studied the language of the region, he preferred to have a quiet, private visit. The pathway, rutted by wagon wheels and pummeled by horse hooves, led to a charming stone bridge. Todd squished out another snapshot. The river gurgled beneath the stonework. He leaned over the bank at the bridge's edge. It was a beautiful river. He stepped onto the overpass, mesmerized by the calm of the early morning.

The river's sound masked the oncoming cart until it was nearly upon him. The driver called out, pulling hard on the reins of the oxcart. Todd had reflexively closed his eyes and put up his hand. He peered up at the driver and his companion, a beautiful, blonde teenage girl wearing a leather apron and a look of alarm.

The man spoke several words in quick succession. Todd searched his memory. With relief, he recognized the words: the oxcart driver was speaking d'oc, the Occitan language of the southern Kingdom of France. Todd was having a hard time putting a sentence together, but he tried to speak the words for friend and greeting.

The man in the cart looked at him sternly. He stood and waved in the direction they had come. He began to speak again. *What's wrong with me?* Todd thought. *I've studied this stuff backwards and forwards. I should know this.*

His head throbbed and his thinking was fuzzy. The only word he caught was "affreux"—*that which troubles life.*

The driver urged his oxen on. Though he seemed concerned for Todd's wellbeing, he was obviously in a hurry. The cart rattled over the bridge and disappeared into the trees on the far side.

Todd turned to head back to the pod. Up ahead, on the road in the direction he needed to go, the pounding of hoof beats thudded along the trail. This must be what the oxcart driver and young woman were fleeing from. He ducked into the tall grass, making himself as flat as he could. The horses approached. When they had come up nearly even with him, they stopped. He could hear voices, though he only caught every third or fourth word. There were at least two riders. Todd carried nothing of value. If they spotted him, they would be disappointed thieves.

One of the riders dismounted. Todd closed his eyes and prayed. When he opened them, a boot rested beside his outstretched hand. He turned his head to look up. The man seemed huge from this angle. He was well-built with a leather jacket and a nasty-looking dagger. He wore his hat at a fetching angle. Its black material formed a frame with his scraggly beard. He said one word, and it was one Todd knew: *stand*. He stood. The man pointed at Todd's clothing and made some comment. Todd missed the specifics, but he recognized the sarcastic tone. The man and his two companions laughed. The others dismounted. Now it was three against one. The pod was probably too far away to outrun these thugs, and Todd was still feeling woozy from the Q-slip.

The first man reached over and searched Todd's pockets. As he did, the cloth began to unravel. He grimaced as he pulled the bioimager from Todd's side pocket. "Lunch?" he asked, dropping the gooey object to the ground.

"Oh," Todd groaned.

"Your boots," the man demanded. Todd took off his shoes. The man held one up, baffled at the design, but since the sole was coming loose, he handed it back to Todd. He stared at Todd's chest, through what was left of his tattered shirt. It was the talisman—Todd's control unit—that had caught his attention. If the guy took it, and if the pod was gone by now, Todd would have no way home.

The thug reached over. Todd stepped back. One of the others stepped behind him, brushing against his back. The message was clear: hold still. The leader grabbed the talisman and yanked it off, breaking the lanyard. He held it up to study it. Todd's hopes sank as the man turned the glittering object in his hand. A grin slowly spread across his dirty face.

"Charmant," he hissed. *Lovely*. He moved his fingers over the object.

Todd searched for the words and finally said, "Be careful. It will…break."

The man looked at him, puzzled. Todd tried again. "Delicate. Dangerous."

He turned to his companions. "Dangerous?" he guffawed. They joined him in his amusement. His thumb dragged across the back of the control device, against the button. Todd thought they were too far to trigger the pod. He was wrong. The man vanished.

26

Dresden Visits the Boss

Dresden remembered with satisfaction the discomfort Rex Berringer had shown during his latest visit to the PI's office. Dresden would not give the same satisfaction to his own boss. "The Ambassador" never called him Tex. That bothered him. He wanted to be on friendly terms with his overseer, but he wondered if the Ambassador was on friendly terms with his own mother. The man was an enigma. Not quite right in the head, and crafty. From the first, Dresden had sensed something wicked about the man.

The Ambassador liked to keep things confidential. It was as if he was censoring his entire life. He was a controlled, self-disciplined, private man. Dresden had always been on a strictly need-to-know footing in the Ambassador's eyes. Still, the two were on close enough terms that Dresden could sometimes incite him to share, but that level of sharing depended entirely on the man's mood, and Dresden still had difficulty judging just where that lay at a given time.

"No, Dresden, it appears that we cannot change the past in a linear way, although we can borrow from it."

"Borrow?" Dresden asked.

The Ambassador paused. "It is always easier to go forward in the timeline than back. Travel in that direction presents no paradoxes. You see, by manipulating elements of the timeline, people or actions, we can change what has not yet happened."

"You're talking about changing history," Dresden said flatly. "Besides, I thought we were doing fine in our recruitment." He was careful to include himself in the Ambassador's group—our recruitment—though at times he preferred to distance himself.

M. Carroll, *Plato's Labyrinth*, Science and Fiction,
https://doi.org/10.1007/978-3-030-91709-8_26

"True, certain factions have been sympathetic to our cause." At first, it seemed as though the Ambassador was finished with the conversation. But after a moment, he met Dresden's gaze. New energy seemed to flow through him. "Prescription drugs are the practical side of my business, but not the side I'm emotionally invested in. As for that other side of my business, global results have not been expansive enough, Dresden. Not nearly fast enough and not the numbers we need to make real change."

Dresden shifted on his feet, trying not to appear uncomfortable as the Ambassador ramped up. The man clenched his hands into fists as he spoke.

"There's something in the wind, you know. Do you feel it, Dresden?"

He nodded noncommittally.

"There's something in the offing that I haven't told you about, too. It's a conference. In South Africa. Very hush-hush."

"I suppose that's why I haven't heard of it."

"It's an international meeting of like-minded people. We'll craft plans and strategies, Dresden." He pounded fist into a palm. "Plans! And strategies! The world's a mess. It needs organization. Rule of law, not anarchy. Unity under strong leadership. A world government that everyone can count on."

Dresden wondered why young, knobby-kneed men in brown shorts flashed across his mind. Could a world government ever engender a free society? It hadn't worked out for Genghis Khan, or Chairman Mao, or Stalin.

"Compassionate leadership, Dresden. That's what's needed."

"Wasn't Hitler compassionate to those around him?"

The boss smiled patiently. "Hitler was passionate. Not *com*passionate. I admire how grand his vision of empire was—how he attempted to form a new world order. Yet those with power need to understand empathy and order in equal measure. For that lesson, we can turn to the most well-disciplined warriors of the human epoch."

"The Swiss? They came up with those army knives. And the cheese, of course."

The Ambassador faltered. "What? No."

"The Ottomans?"

"No, no. You see, Dresden, we need a disciplined police force, not a conquering army. There was a problem with all the other empires, all those proud conquerors. Do you know what it was?"

So many retorts flashed across Dresden's mind, but none of them would send the dialogue in a positive direction. He let the Ambassador go on without reply.

"Hitler's forces, and those of Genghis Kahn and various kings and emperors, followed a person first and foremost. Certainly there were those who shared the vision, whatever it may have been. But the majority followed the leader. Hero worship. The Romans, on the other hand, had it right: they pledged allegiance to the Caesar not because of who he was, but because of the ideals he stood for. The Romans nearly conquered the world, and they may well have, if their minds and hearts had only remained pure of motive. The world was not ready for the Roman Empire then, but the world needs it now. We will make use of their standing armies, the Roman legions. We will avail ourselves of the cream of the crop, the Praetorian Guard. Especially if we go back early, perhaps to the time of Caesar Augustus or so. This will be a new army of the global world. Not a dark and secret police squad, but an open organization trained to keep society orderly and serene."

Dresden wondered what the Jews of the first century would think of the Ambassador's view of the beneficent Roman occupation. "Why not conscript your global police from among modern forces?"

The Ambassador began shaking his head as soon as Dresden started his sentence. "Armies can be polarizing. People hold grudges for generations. Our Roman legions will inspire as much loyalty as they follow. They're far enough removed from local politics that nobody will have a grudge against them. On the contrary, their history has been heavily romanticized." The Ambassador swept his hand across the room, toward a red banner framed by Corinthian columns. "They will be seen as legends; heroes in their own right. People will listen to them over the petty armies of today."

"Gotcha," Dresden mumbled.

"And there's another aspect to our plan, a vital one. It's not just about importing disciplined forces here. It's about their built-in leadership. And it's about recruiting brilliant strategists."

"You have someone specific in mind," Dresden guessed.

"Oh, I do indeed. I do indeed. Lucius Seius Strabo. He was a prefect of the Roman Imperial Guard, a high official and personal bodyguard to both Augustus and Tiberius. He was just the kind of tactician we will need. Brilliant mind—one of the greatest in history. Any of our recruits from the era will follow him, and follow for the right reasons. Under him and the guidance of the Praetorians, the world enjoyed the Pax Romana, peace throughout the empire. And that's what we will have. The Pax Primus. The Pax *Imperium*."

But peace comes at a cost, thought Dresden. The Ambassador flipped his hand dramatically. "I suppose you could say we are now in the import/export business. And Mr. Dresden, your skillset will be needed. As Lucius will know nothing of technology or politics, you can bring him up to speed. And his background, coming from ancient Rome, will actually give him the advantage of thinking in ways new to us. He'll think outside of the box, as it were."

For the first time, Dresden felt true alarm welling up in his chest. He tried to sound merely curious. "Is this new part of the plan—this importing of troops—imminent? How will we get green cards for all of them?"

The Ambassador dropped his voice a pitch. He spoke softly, almost menacing. "If we want the Imperium's wave to break before mid-century, we must move now. My engineers are preparing our first temporal capsule for a test transport."

Temporal capsule? They even had a name for it.

The Ambassador leaned toward Dresden. They locked eyes. "The world is suffering, teetering on the verge of anarchy. Every day the news is worse. It cannot wait. Understood?"

Dresden did understand, but didn't like it. Shuffling the quantum chessboard did not sound like a good idea. Dresden had always been lousy at the game, but he was familiar enough with it to know that chess only worked when the players followed the rules. How many were they about to break?

He couldn't help but ask, "Why would these soldiers climb voluntarily into your contraption?"

"Timing. The warriors we take will be plucked from their encampments on the eve before a battle they know they are doomed to lose: the Battle of the Teutoburg Forest, 9 AD. The event was such a catastrophe for them that the Romans had their own term for it: the Varian Disaster." The Ambassador laughed at this and continued. "So, our recruits will be faced with the choice of slaughter or another chance in another time. And in this way, they won't be missed on the timeline."

What if the very act of plucking these armies from Teutoburg was the actual cause of the Roman defeat? Dresden wondered. And flooding the third millennium with a bunch of soldiers from the first did not sound as easy as the Ambassador made it seem. On the other hand, he was paid far too well to jump ship just yet. He could always bail if things got out of hand. Meanwhile, he would turn his attention to more pressing matters. The Ambassador needed to know how ChronoCorp had progressed in its research. Had they solved the same problems that the Ambassador's people had? Did they know more?

Perhaps Dresden needed to tell the PI more than he had. The man was not exactly incompetent, but he certainly was slow. Perhaps Berringer would surprise Dresden, and his next visit would yield the files they needed. Then maybe, just maybe, ChronoCorp's work would put the Primus Imperium over the top and on its way to a multidimensional army. Dresden shuddered at the idea.

Something wicked this way comes.

27

Homecomings and Goings

Katya tapped her foot. Todd had been gone longer than architecture five called for. She looked across the room at the empty pods and the three talismans hanging from the wall. Next to them was an empty spot for the fourth unit, as if to emphasize their missing colleague.

"Power's coming up," one of the techs called out.

Katya's monitor flashed as the power spiked. There was always a subtle change when a traveler arrived. The pods had no lights on them, no horns to announce someone's return, but in just the right light, there was a faint shimmer, and the feeling of a presence that hadn't been there before. Katya had that feeling now.

Todd didn't open the hatch. Usually he sprang from the pod in triumph. Now, there was only silence.

Xavier stepped over to the pod. The glass was fogged, obscuring the figure within. Xavier opened the hatch, then jumped back. Katya rushed over, prepared to perform CPR.

"Do we need medical?" she cried out, but then she froze.

The man inside sat up, still clenching the talisman in his hand. He looked around… and fainted.

"I'll be right back. Just keep him in there!" Xavier rushed out the door.

Katya quietly closed the lid again, praying that the intruder would remain unconscious.

Xavier returned with a pistol.

"Where did you get that thing?" Katya snapped.

"Keep it in my truck. I have a license."

"Yeah, but what do you plan to do with it?"

© The Author(s), under exclusive license to Springer Nature Switzerland AG 2021
M. Carroll, *Plato's Labyrinth*, Science and Fiction,
https://doi.org/10.1007/978-3-030-91709-8_27

He held it up, carefully. "Intimidate, if I need to. I'm going to escort our unwanted friend back to his time and find Todd while I'm at it."

"Who knows where he's gotten to?"

"Some time has elapsed, time enough for Todd to have gone off and gotten into some kind of trouble. If we set our target to return a few minutes earlier than when this guy departed, say, five minutes after Todd arrived back there, we can be sure he'll still be near the pod."

"But that's changing the future, even though you're doing it in the past. You would be changing Todd's future."

"Just a few minutes of it."

"Xavier, we haven't done this before. Not a tandem with people. I'm worried that—"

"The math says it will be fine."

"You're the boss," Katya said nervously.

"Every time you say that, it sounds more like 'It's your funeral.' Let's just try it." He grabbed a talisman from the wall and stepped to the second pod.

Katya called on her headset. "Are power levels back up enough for two more transports? Yes, a tandem… We're sending two out, and two will return." *But not the same two, hopefully.* She handed Todd's talisman to Xavier. "We're good to go."

Xavier climbed in and shut the lid. Katya called to Brianne, who was stationed behind a monitor at the other side of the room. "Did you get all that, as far as the timing of this thing?"

"Got it logged in and confirmed. Computer is ready to accept."

Katya grimaced and said, "Let's go."

The room hummed, the lights flickered—they would have to get that fixed one day—and the coffin-like pods emptied.

*   *   *

Xavier knew he had arrived…somewhere. The throbbing in his head told him so. The pain and confusion were so intense that at first, he couldn't remember where he was supposed to be. But then he heard the voices. They were speaking French. And it all came back to him.

He sat up quickly, reeled for a moment, and rolled out of the pod onto all fours. His breathing came in rasps. He stood unsteadily, grabbed the gun and the talismans, and wheeled around to see the other pod. It was still closed. The voices were coming from close proximity, somewhere behind a grassy berm along a dirt road. Xavier held the gun in front of himself, barrel skyward. He didn't want any accidents causing quantum weirdness in the timeline—his or theirs. But he figured one shot in the air would send anybody from this era running.

He started down the road and heard a line of his native tongue. "Look, I have nothing else, and I'm sure your friend is fine." This was followed by a halting stream of French, met by a torrent of French from two others. Xavier rushed toward the sound and came upon Todd, facing two large men with equally large knives. The sunlight flashed on one of the blades.

Two of them and Todd—but no sign of the one who had visited ChronoCorp.

Todd spotted him. "Xavier, am I happy to see you!"

"Glad to make it to your party. Let's get back to the pods. I've got your talisman."

"The other guy?" Todd stared at the gun. "You didn't—"

"No, no. He's sleeping inside the other pod, last I checked. Let's roll."

Todd wanted to tell him of his discovery—that even being close to a pod was good enough with the talisman. But Xavier was already on the move. The two thieves had lost focus, relaxing their weapons. Now, one of them turned toward Todd. Xavier called out and held his gun aloft. The man didn't recognize it as a weapon, and continued toward Todd.

The Glock 19 was a superb firearm. In the four years that Xavier had owned it, it had never disappointed. He pulled the trigger. Nothing happened. He said, "Run!"

They did. They reached the pods just as the thief in the first pod fled across the field. Todd and Xavier plopped into their pods. Slamming the doors, they both triggered their talismans, leaving medieval France behind.

28

Platypus Panoramas

The bioimager was in bad shape, but Xavier didn't care. He was jabbing the table with his index finger. "We successfully traveled concurrently in two pods. Two pods, Todd. That's a record, and it means we're ready for Thera."

Todd kept his eyes on the bioimager. It was encrusted with dirt and grass, and it was twitching.

Emil carefully brushed it off. "I don't know, guys. The thing's not happy, I can tell you."

"Is it salvageable?" Todd asked.

Emil plugged it into his computer port. In moments, a beautiful image of the river Garonne appeared. The three cheered.

"Is your gun functioning again?" Emil asked Xavier.

"Looks like new, but I decided not to try it in here."

"Probably a good call," Todd mused.

"Odd that it didn't work back there, but I should have known. Guess it's the same thing that goes on with all our technology. Mechanical things just don't translate across the continuum."

"I feel like we dodged more than one bullet on this one," Todd said. "I escaped those middle-ages-hoodlums by the skin of my teeth."

"You didn't stick to the plan," Xavier said. "Thera will be the same case. No interaction with the locals. Got it?" He said the last with a smile, but his tone was firm.

"You got it, boss. And we learned a valuable lesson: we don't have to actually be next to the pod's landing site when we call it."

M. Carroll, *Plato's Labyrinth*, Science and Fiction,
https://doi.org/10.1007/978-3-030-91709-8_28

Xavier frowned. "I wouldn't push that one, though. We don't know what kind of range our system has. Better to try to get back to the point of arrival. Let's not get anybody stranded."

Like we almost did, Todd thought. But he didn't share the thought; there was something more urgent on his mind. "So, Boss, I did some digging into the Primus Imperium."

"Did you have to dig far?"

"And deep. They're pretty cagey. Years ago, they started out as a right-wing bunch of survivalists, but they became associated with a respectable pharmaceutical research company, and it seems they've taken a turn to the sciences. Lots of classified research going on there. I hit a few dead ends, but then broke into their monthly energy bills. It turns out they're using spurts of electricity on a fairly regular basis, like they've got an atomic collider or something."

"Really?" Xavier scoffed.

"Well, okay, not that much energy, but a lot more than an office building has any right to use."

"Like ChronoCorp?"

"Yeah, that's kind of what I was thinking."

"Maybe we should look into this further. But after Thera. Take a day or two off first."

Todd's grin was all Xavier needed to confirm the wisdom of his decision.

* * *

Mila had called and texted and emailed. It was getting harder and harder to put her off. Time for a phone call. Even as the line was ringing, Katya realized that part of her urge to call Mila van Dijk was simply frustration. Xavier was being impulsive; he was a moving target. Moving targets were hard to arrange meetings with. The only hope was to associate Mila's appearance with an upcoming event.

Katya regretted the call as soon as the words were out of her mouth. "Mila, we're doing our first attempted run at Thera, and it's next week."

"Any chance of me getting in there to observe?" van Dijk asked.

Katya felt the anxiety tighten her throat. She was having second thoughts. "Oh, that might not be advisable. We don't know what to expect, and there is some evidence to suggest that we won't be able to hit our target on the first try. Besides, Xavier will be so distracted that you won't get to talk much, right?"

"I don't mind. I'd like to see him in action."

Unreservedly lovestruck, Katya thought. "How about this: after our next Q-slip, I'll arrange a meeting between you and Xavier, just the two of you. You as philanthropist, he as grantee. Will that work?"

"Absolutely, my dear. But if you are successful, what is our next step?"

Katya was uncomfortable with van Dijk's use of the inclusive *our*.

"If we hit Thera when and where we target, we need to go back a second time. Proof of concept. Then Xavier wants to hold a press conference ASAP."

"Even before you publish?"

"Rest assured, those papers will come. And soon. We're hoping to present at a big archeology conference in a few months."

"The San Francisco meeting?"

"That's the one. And assuming that goes well, we'll do another presentation at an archeology summit a few months later. At any rate, I need to go."

"Katya, thank you for keeping in touch. It means a lot to this old philanthropist."

"You're welcome, Mila." Love was in the air.

29

1939, Outskirts of Berlin

The last thing the travelers saw as the flash began was the face of the Ambassador, with the big Texan Dresden on one side of him and a tech on the other. The Transference Theater was dim in the same way as a romantic restaurant, with not quite enough light to read a menu. Fred MacMurtry supposed that was the reason for candles. It was the flash of their own light—the light of the temporal capsule they had strapped themselves into—that revealed the observers through the porthole with their excited and nervous faces.

Strange, the places the mind took you under stress. Fred was thinking about romantic Italian restaurants when he was on the verge of the greatest expedition ever taken by humankind, a voyage not across the globe or between planets, but rather outside of the known universe. Marco Polo, Shackleton, Hillary, Earhart, Armstrong—all had traversed the height, depth, and width of reality. But Fred was about to traverse the fourth dimension: time. He hoped he would end up in one piece, unlike the first engineer.

He looked to his colleague, an able engineer named Joslyn. Her presence gave him a feeling of security, even in the bizarre glow coming from outside. The light from the initial flash faded. It seemed to swirl around them, turning into purples, then a mélange of browns, grays, blues, and greens. Gravity seemed to falter, as if the temporal capsule bobbed upon the waves of a stormy sea. Those ribbons of color slowed, melding into ground, trees, sky. Fred vomited.

* * *

© The Author(s), under exclusive license to Springer Nature Switzerland AG 2021
M. Carroll, *Plato's Labyrinth*, Science and Fiction,
https://doi.org/10.1007/978-3-030-91709-8_29

Their capsule sat on a gravel surface, a kind of courtyard surrounded by a low concrete wall. Judging by the surrounding technology and the signs written in German, they had hit their test target perfectly. Ill-fitting red banners hung from the top parapet, some dragging sloppily on the ground as if the entire event—whatever it was—had been a last-minute affair. They had arrived in the midst of an assembly. The gathering was well attended. Rows of chairs were filled to capacity, and people stood along the back wall to the capsule's left. Fortunately, all of them appeared riveted to the speaker at the front, a young man behind a tippy lectern. He gesticulated wildly about something as the crowd applauded.

Fred glanced around. Loudspeakers—large metal horns—stood atop posts at either side of the podium, bellowing a scratchy version of the orator's voice. People, most of them men in khaki, waved flags, banners, and scarves. But what really got Fred's attention was the tough guy approaching them from the right. The uniformed man pulled out a menacing sidearm. To Fred, the guy looked like a *shoot-first-and-ask-questions-later* sort. Joslyn saw him, too.

"Well," Joslyn said, "it worked. Let's get out of here."

Fred tapped a touchscreen. The landscape fractured, as if God himself had taken a saber and flayed it into strips. The flash came again.

* * *

The Ambassador was beside himself, fidgeting and pacing as the minutes stretched on. He kept smiling at Dresden, something that made the Texan uneasy. He'd never seen the Ambassador's teeth. The techs watched their screens. The theater remained empty, except for a much longer capsule than the one that was currently in use. Dresden knew what the bus-shaped chamber was without asking: the Ambassador's troop pod.

The Ambassador's smile remained, but it couldn't mask his nervousness. *Where could they be?* "Perhaps they're conversing with some of the people there," he said to a nearby technician.

The tech shook his head. "No worries. We think time will pass differently for our travelers than it does for us. Patience is the order of the day, Sir."

The Ambassador was too distracted to argue. He turned to Dresden. "Where would you go? Once we get the hang of this, get this world stabilized, what time would you go back to visit?"

On the inside, Dresden was appalled. The man was speaking as if this fearsome technology was a ride at Disneyland. *Pirates of the Caribbean* on steroids. "I'm not sure I'd go anywhere. I have enough troubles with my own time."

One of the screens beeped. The lights in the room flickered, dimmed. All the computer monitors blanked, then gradually came back. As the lights flared on again, the cabin stood at the center of the Transference Theater. Two techs stepped carefully to the small compartment and unsealed the hatch. They froze.

"Well?" the Ambassador called. "How are they?"

One tech leaned in. His two colleagues lay on the couches of the capsule, as wrinkled and desiccated as Egyptian mummies. Their eyes were gone, leaving cave-like hollows to stare from their yellowed faces. Joslyn's wizened arm was bent at the elbow, her hand seeming to shelter her face.

The other tech straightened, turned toward the Ambassador, and said, "They're dead, sir."

With a grunt, Dresden turned and headed for the exit. *Still a few kinks to work out.*

30

A New, Very Ancient World

The Transport Arena buzzed with energy, most of it coming not from transformers or capacitors or cables, but rather from the technicians crammed into the room and in offices down the hallway. Katya Joshi stood at her console in the Arena, and as commander for this Q-slip, she directed the rooms like a concertmaster. She spoke into her headset as Xavier and Todd climbed into their pods like the old astronauts shimmying into their Gemini capsules.

"Mark One," she said, as she had several times before. This time around, though, the excitement had an edge of apprehension. She could feel it. Maybe it was because two people were going. Maybe it was because she felt it was all a step too far. Where was the quality control? What happened to the methodical scientist Xavier used to be, conservatively testing and retesting before making a new move? He was a different person these days, and not one she wanted to get too close to.

"Mark One confirmed," a voice intoned through her headset.

Xavier closed his lid first. Inside the glass top, he held up his bioimager toward Katya, as if proposing a toast. Todd followed a few moments later, carrying a second bioimager, a present from Emil to the team. *Two disgusting things are better than one,* Xavier had commented.

"Have fun, boys," Katya called through the acrylic lids. "And may the wind be at your backs."

Todd gave Katya a thumbs-up through the clear cover.

"Mark Two," she declared. "Our explorers are tucked into their little beds."

"Mark Three," called another tech. "Power up."

Behind the ChronoCorp building, the structure that resembled a cell tower came to life within. A spiral of powerful lasers blazed, their corkscrewing light

© The Author(s), under exclusive license to Springer Nature Switzerland AG 2021
M. Carroll, *Plato's Labyrinth*, Science and Fiction,
https://doi.org/10.1007/978-3-030-91709-8_30

pulling space and time with them. Slowly, gradually, the very fabric of space began to fold in upon itself, like a miniature black hole moving through the columnar structure.

"Mark Three confirmed," an engineer called out. "Lasers at full."

Typically, the departure of the pods was somewhat anticlimactic. No bright lights or bizarre quantum warping of the room, just a sizzling sound, a pop, and then silence. There one moment and gone the next. But apparently, this was no ordinary trip. Katya waited. The sound certainly came, but it was harsher than usual, like a goliath's steak burning on an over-heated grill. And then came the surprise: a dazzling flash that filled the room with purple light. Katya was temporarily blinded. As soon as she could make out her monitor, she looked to the pods. Todd and Xavier were gone. She tapped her headset. "What the heck was that? Power surge?"

No one responded.

"Emil, anything?"

After a moment's pause, Emil came online. "That was no power surge. Everything is still stable. Lasers powering down smoothly."

The phone rang, an outside call.

"I thought these lines were supposed to be closed off during all Q-slips!" Katya roared.

Brianne, posted at a corner workstation, cringed. "Sorry, boss. It was on my list, but I got busy with, you know, this." The new hire waved her hand at the room around her.

"Would you like to *get it*?" Katya hissed. It was the adrenalin, coming down. "Sorry."

The tech picked up the line and said, "It's for you. A Dr. Glenn?"

Completely unbelievable. Some people had a gift for showing up at the wrong time. She didn't want Bradley Glenn infringing on this part of her life. Work was a place she was sure of, a refuge, and he was encroaching on it. But she couldn't snub him, even at this critical moment. Glenn was too valuable an asset—and too dangerous of a threat—to ignore. She took the call. As she did, she realized she had balled her hands into fists.

She took in a breath, relaxed, and said, "Brad, we're in the middle of a test right now. Can I get back to you later?"

"Sure, of course."

Good. She could make this call very short. Brad added, "I want to come up and take you to dinner."

This stopped her short. It was a long way from the AMNS to Fort Collins, Colorado. "That's a lot of miles, even with frequent flyer points."

"There's a conference in Denver I'll be speaking at. Thought I could come up. So why don't you give me a call tonight. I'll be up late."

"Okay." She closed the connection without saying goodbye.

"Weird," Katya said aloud. She looked up. All eyes in the room were on her. "Sorry about that. Now, where were we?"

Emil said, "Ancient Thera, I think."

*   *   *

Todd realized immediately that something was off. His Q-slip to medieval France had been more disorienting than the shorter experiments, but this one took the cake. He was nauseated, and the pain in his head was excruciating. He saw so many purple spots that he closed his eyes until they cleared.

Finally, he popped the lid and looked up. Overhead, a pair of pelicans sailed the skies. He could smell the saltwater of the Aegean. He put his hand to his chest and felt the reassuring lump of the talisman device beneath his shirt. The ghost pod had materialized in a small grove of trees against a bank, sheltering it from a village or city behind. Whatever was back there wasn't visible to him, but he could hear the lively sounds of civilization. He saw no pathway, so the site must be safe, for now.

He looked around. He could see the other pod, but where was Xavier? Stepping gingerly onto the soil of ancient Thera—one small step for a man— he made a beeline to the other pod. Through the glass, he could see Xavier. The poor guy was rubbing his temples and grimacing, his long legs folded against the side of the chamber. Todd tapped on the glass. Xavier opened his eyes.

Todd couldn't help but grin. "We made it, Boss!"

Xavier unsealed the pod and sat up. He wavered like a child's top on a table, nearly falling over. He let out a groan. "I hope they've got something stronger than acetaminophen here."

Some kind of pain reliever sounded good to Todd, too. But at that moment, a sound distracted him. Floating on the Theran breeze was the Minoan music of the dresses. The music played, a symphony with a hundred members, all hemmed in those light copper disks that sang the song of wind chimes. It was just as Mila van Dijk had imagined, the sultry smells of roasting seafood and vegetables, the aroma of sulfur coming from the city fountains or from Thera's guts, the bracing Aegean zephyrs. But what Mila had not anticipated, what Todd had failed to imagine, was the laughter, the bustling conversation in a foreign, long-dead tongue. Although he couldn't tell the content, he could tell the context: a market or town square, the back-and-forth banter of negotiation and tête-à-tête. He wanted to see, but he knew caution was the order of the day.

Xavier swung his legs out and stood, wincing. One of his sleeves tore at the shoulder, then slid down his arm, reminding Todd of a snake shedding its skin. Clearing his throat, Xavier croaked, "We'd better get moving. We've got to at least get a glimpse." He held up the organ-like bioimager. It let out a soft burp.

"Which reminds me," Todd said, "Did anyone remember to feed the cameras?"

"Emil did the I.V. this morning. The next one of these they make should have a mouth."

Todd grimaced. "I wouldn't give these things any teeth."

The two explorers limped unsteadily, gaining the crest of the ridge behind the pods. The sounds of the town came from up and to the left, but the ocean lay to the right. The Aegean waters glimmered like a gemstone, rich green and ultramarine. The stony bottom stretched far out to sea, losing itself in the azure rise and fall of gentle waves. A craggy shoreline formed a rocky crescent. Its indentation into the land served as a sheltered cove, dotted with single-masted ships. Stevedores shuttled back and forth between the moored boats and a row of low-lying stone structures. Stretching across the waters toward the horizon, a processions of stone jutted from the sea in parallel lines. No, Todd thought: not lines, but rather concentric circles surrounding the island. Just as Plato had described his Atlantis.

"Shall we get a closer peek?" Todd asked.

Xavier blinked his bloodshot eyes, meeting Todd's gaze as if coming out of a coma. "Yes, of course."

Todd wondered if Xavier could make it down the hill and back again. "You want to stay here, Boss? Maybe hang back and keep a lookout?"

Suddenly energized, Xavier said, "Not on your life. We've worked so long, so hard. Besides, if we're going to stay very long, we need to find some clothes."

They were finally getting their sea legs, feeling steadier. Descending the hill, the terrain leveled out. Mediterranean pines rose in candelabras of gray-green. The duo climbed down to a rocky outcrop. As they settled into a spot where they could watch the goings-on unobserved, Todd looked up the hill behind them. There, looming above all they surveyed, rose the cone of the mighty stratovolcano in Thera's central bay. Trees marched up its dark slopes, verdant against the sulky volcanic soil. Red roofed buildings similar to the warehouses at the shoreline were strewn across the summit, and a tapestry of cultivated farmlands cut terraces into the mountain's face. A collar of gray volcanic rock ringed the summit where no trees would grow, and no one would want to build a house. They could smell the fumes even from here. Farther downhill, rivers of glistening water meandered down the slopes. At the top, smoke drifted from the pinnacle. The plume seemed so serene, so graceful. But Todd knew that in a short number of months or years—they were unsure of their arrival date—25 cubic miles of the mountain would vaporize in an apocalyptic cataclysm. Here, too, lay the answer to the enigma of Thera's form. The volcano was no humble cone in a lagoon. It was massive. Whether it formed a bridge across the circle of Thera, he couldn't tell from here.

Todd heard a squish, the unmistakable shuttering of a snapshot by the bio-imager. Xavier wore a grimace. "It's sticky."

"Yeah, I know. But if it works, all the messiness will be worth it. Did you get the ships?"

"And the warehouse down there. I thought you should take a few detail views of the architecture."

"Sure thing." Todd took the pulsing lump from his own pocket, careful not to drop the slimy object. A dirty camera would not do. Todd pointed its eye-like orifice forward and shuttered the structures, their roofs and gutters, the lintels and sills. Most of the buildings here were simple blocks with small clerestory windows and large doors facing toward the main road leading to the wooden docks.

"Those must be storehouses, Todd reasoned.

"Why?" Xavier asked with the curious tone of a true researcher.

"Look at the arrangement: a main causeway leading from the shore to the courtyard, and all these big buildings are oriented with their wide doors toward the access to the ocean. Clerestory windows to keep the interior cool. Maybe granaries?"

"Wish we could see in," Xavier said, but his tone sounded more tired than interested.

A row of pithoi, much like the urns that Mila van Dijk had in her museum, lined a wall just across a cleared yard. "I'd love to get closer to those. Get a couple detail shots."

Xavier squinted in their direction. "Let's make it fast. And then we go. We've got plenty to take home already."

Surreptitiously, they crossed the open area. Xavier stood with his back plastered against the wall, scanning the scene. Todd took an image, then another. "These are rugged pots," Todd whispered, "but these on the end, these are amazing. Look at the glazes…"

His voice died as he stared at the face of a Minoan merchant peering around the large vase.

31

Big Bruisers

Dresden rarely wore sunglasses, but he was wearing a pair now, as if to shield himself against the coming tirade. He could tell this encounter with the Ambassador would be different from others. Dresden had been summoned to the warehouse. But the Ambassador was not in his office.

"The Ambassador is at the north facility," an attendant said, each word carefully clipped. "He asked us to bring you to him there."

What north facility?

The attendant—a fireplug of a man—walked Dresden out to a large grey SUV, a vehicle almost as menacing as Dresden's own. He stationed himself by the driver's door. A second human guard dog told Dredsen to raise his arms. Dresden knew the drill, of course, but he had never liked being frisked.

'Hey, none of that," the driver said flatly. He put the palm of his hand on Dresden's arm and gently pushed down.

"It's standard procedure," the guard snapped.

"Our passenger is not a standard case."

"Don't blame me if you end up with a hole in the back of your head," he grumbled, climbing into the back seat. "You've got enough of them as it is."

The assistant had a suspicious bulge on the belt line at about the five o'clock position. It looked like a good fit for one of the Glocks, or maybe a Sig Sauer. Dresden didn't want to find out, although he did have a couple of his own. He kept his three-inch Smith & Wesson 357 Magnum on the belt line at the back, easy to grasp, compact but powerful. He liked revolvers. They had history. But the bigger ones were also a bit obvious on the street. He carried another Smith & Wesson, this one a .38 Special in an ankle holster. Whenever he thought about his weapons, he remembered the Jean Shepard story about

© The Author(s), under exclusive license to Springer Nature Switzerland AG 2021
M. Carroll, *Plato's Labyrinth*, Science and Fiction,
https://doi.org/10.1007/978-3-030-91709-8_31

Ralphie's Red Ryder bb gun with "the compass in the stock and the thing that tells time." It was a good story for a kinder age.

Dresden knew the other guy was packing. He couldn't see what, but he bet it didn't have a compass in the stock. The man appeared to be wearing a shoulder holster. Shoulder holsters weren't a popular fit for a quick draw; they were harder to access than one at the belt, and the pull usually swept a part of the body, or any close bystanders. But in a car, a shoulder holster kept the gun clear for easy access. Dresden assumed his companions were more concerned with the practical.

The two-lane road out of town followed the thundering Cache la Poudre River, swollen from recent rains and melting snow farther up into the Rockies. Blue columbines and wild raspberries grew along the water's edge. Bastions of granite towered like titans at attention, waiting for inspection. Within their crags, spruces and pines rose in parades, accented here and there by the speckled black and white trunks of aspen. It might have been an enjoyable ride, if Dresden could only relax.

The SUV slowed at a point that looked, to him, like any other in the miles behind them. As the car pivoted right, a NO TRESPASSING sign appeared, angry red letters on a black background. The trees parted onto an entrance of packed gravel, hidden from the road by the dense foliage. A formidable gate topped with razor wire blocked the way. If the sign didn't put off any hikers, the sight of the gate certainly would. Notices on the fencing declared HIGH VOLTAGE. DANGER.

"Looks like a pretty serious place," Dresden said, deadpan.

"You have no idea," the driver muttered. He punched a button on his dashboard and the gate swung open. They proceeded inside. The road split into two paths walled by dense trees. They took the one to the right. The boulevard narrowed into a one-lane affair with essentially no shoulder. They reached a clearing with an open parking lot. On one end, an observation tower oversaw the entire area. Dresden had visions of Stalag 17. He wondered what kind of guard towers the ancient Romans had built all those centuries ago.

The barricade backing the lot was not flimsy wire, but solid masonry. The stonework was utilitarian; no landscaper had crafted this fortress, unless the architect had been into mid-century penitentiary. The three made their way to an entry bracketed by cameras. The driver held up an ID and a latch clicked. As an armed guard let them in, a helicopter passed low overhead, traveling in the same direction they were going. It disappeared beyond the treetops, but Dresden could still hear the rhythmic thudding of its blades.

Beyond the stand of trees rose the main building itself, done in elegant concrete block. Oddly, the main entry sported incongruous doorposts of fluted columns right out of the Roman Forum. 30-foot-tall banners hung from the roof, draping the front in a parade of colors. Dresden recognized some of the symbolism; it was hard to miss. To the far left hung a sail emblazoned with blue chariots and distinctive rounded chevrons. Dresden guessed it stood for the armies of Genghis Khan. Beside it flew the flag of the later Ottomans, red with white crescent and star. Next to it spread the banner of Mussolini's Italy, and to its right, the rising sun of the Empire of the Sun, both circa 1940. Then came the black German swastika encircled in white against a vibrant, blood-red background. The union jack of the United Kingdom followed. American and Soviet flags of the Cold War came next. These seemed to Dresden a controversial choice at best, but he supposed the Ambassador could decorate his building however he wanted. Finally, edging the entryway, the undulating iron eagle of the Roman Empire. On the other side of the door, an identical banner began a mirror image of the banners to their left. The theme was unmistakable: empire. World dominion. Dresden began to wonder about what kind of *Pax* the *Pax Imperium* would embody.

Once inside, the driver held his wrist to his mouth and spoke a few unintelligible words. He dropped his arm and said, "This way."

Dresden followed down a long corridor that reminded him of a surgeon's theater, polished to a shine on floors, walls, and ceilings. A succession of doors—all securely closed—stepped up in number. If the numbering system followed anything conventional, the facility must have had dozens of rooms, at least. They reached a double door at the end. Passing through, Dresden found himself on a high balcony, a catwalk that overlooked a huge warehouse-like area. Across the concrete floor, dozens of tables stood in long rows, lined with plastic chairs. The place reminded him of a school cafeteria. The room could seat hundreds of people at once.

"What is this place?" he asked, not taking his eyes from the sprawling room.

"Mess hall. For the future troops."

After a moment of silence, the driver cleared his throat.

Dresden was so entranced with the place that he almost apologized. But apologies were not in his normal lexicon, and he regained his composure as they reached a glass-walled conference room. Inside, the Ambassador was fuming behind a table. As Dresden entered, he stood to look out a small window.

"We've got to face facts, Dresden: our quantum cabin has now killed three people. Three valuable technicians. I have spoken of our need to obtain details—and not just surface-level details, but...*detailed* details," he shrugged at his own inability to choose an appropriate adjective, "on how ChronoCorp is doing business. Have they actually traveled yet? Certainly their power usage would argue that they have. Were they successful? Did their people come back alive?" The Ambassador paused, and Dresden let him. When someone says something like "detailed details," it's best to give them a little space.

The Ambassador continued in a higher pitch. "The next step, of course, is our chronometric coaches to transport a dozen passengers at a time. And we're certainly not near to that, yet. Mr. Dresden, you—*personally*—need to inspire our private investigator to get those files. If this doesn't work by the end of the month, if he can't access those files for us, or if those files are of no help, get rid of him."

"You mean…"

The Ambassador waved his hand as if swatting a fly away. "Stop being so melodramatic, Dresden. Pay him a good severance fee and bid him farewell. It's about time we made use of our illustrious infiltrator anyway."

Dresden noticed the alliteration and assumed the Ambassador was trying to be clever. But the two words were antonyms. If a stealthy infiltrator became illustrious and well-known, his work would be of no use.

Or, in this case, *her* work, because the Ambassador had told him the identity of the mole, but Dresden wouldn't be telling anybody in the near future.

He wasn't sure how to safely phrase the question burning in his mind, but he wanted to know badly enough that he forged on. "What will we do with that knowledge? What happens when we obtain the information we need to build our global police force?"

"Our utopia, you mean," the Ambassador corrected him. "What will we do? What we must, for the sake of a peaceful world. We try to forge an alliance with our main competition, ChronoCorp."

"And if they won't play?" Dresden asked.

"Then we have no choice but to destroy ChronoCorp and anything related to it."

32

Welcoming Committee

The burly Minoan wore a kilt-like affair around his waist, heavy boots, and an open vest. A colorful headband ringed his mane of wavy black hair in a thin strip. He spoke. His tone was neutral. Todd looked desperately at Xavier. Surely something in his Aegean studies would help? But Xavier only shrugged. So much for his "Minoan A."

Two more men and a woman carrying a clay tablet rounded the corner of the wall and stopped short. The big guy closest to them gestured in their direction and addressed his cohorts. As he did so, Todd's belt disintegrated. His pants fell off. This event triggered energetic conversation amongst the Minoan merchants. The group seemed to defer to the woman, who was apparently in charge. She wore a beautiful bodice-jacket around the shoulders. Her chest was as exposed as those of the men. She stared at Todd's disintegrating wardrobe, a faint smile tugging at the corners of her lips. She said something else to her cohorts.

Todd looked down. "This is going to get downright embarrassing."

Xavier's face lit up. "Foreigner. I recognize the word! Foreigner, or perhaps visitor."

"That's a whole lot better than 'intruder' or 'enemy who must be skewered,'" Todd said under his breath.

"Certainly not enemies," Xavier said, nodding toward one of the men. "Look." The young Minoan had removed his kilt. He handed it to Todd, waving his hand toward the fragmenting pile of cloth around Todd's feet. Another handed his own skirt to Xavier, and still a third doffed his scant waistcoat, shoving it into Todd's hands. The material was light and cool. One of the men disappeared around the corner and returned quickly with mix-and-match items of clothing for those who had donated theirs to the cause. Todd made

© The Author(s), under exclusive license to Springer Nature Switzerland AG 2021
M. Carroll, *Plato's Labyrinth*, Science and Fiction,
https://doi.org/10.1007/978-3-030-91709-8_32

note of this: on any return trips, it helped to know where the inhabitants kept the extra wardrobe.

Todd struggled with the ancient fashion. He couldn't seem to get the garb to stay on his shoulder. The man who had given him the shirt reached over. Speaking a few unintelligible words, he pulled the cloth up and tucked the slack into Todd's waistband. As he did, Todd noticed the woman again. She had glanced over her shoulder at him, and in that brief moment a smile had shadowed her lips.

When Xavier and Todd were properly attired, Todd slipped the imager into the folds of his clothing. He hoped Xavier still had his. The merchants surrounded them, gently ushering them toward the city. But why?

Are we welcome guests, Todd wondered, *or lambs to the slaughter?*

The Minoan merchants led Todd and Xavier down the cobblestone streets of the village. The rooftops had been crafted of red clay tile and beaten copper, reminiscent of Italian villas Todd had seen. The walls stood in a polychrome of blonde, white, rust and black, all colored pumice quarried from the hide of the mighty Thera. Niches inset into the walls contained vases or small clay figures. And as Mila van Dijk had surmised, they passed spring-fed fountains, glistening in the sun and smelling of rotten eggs. They crossed a town square where a gaggle of young men and women set up tents and booths. The garlands and empty tables spoke of something not permanent, perhaps a festival.

The people wore a variety of costuming. Men and women alike wore heavy belts that accentuated their waistlines, and bodice-jackets, although the men more often wore nothing above the waist. Heavy, thigh-length aprons, ankle-length tunics and flounced skirts abounded. Many of the women and girls wore diaphanous veils or head coverings, and the women wore golden jewelry: necklaces, hoop earrings, bracelets, anklets and collars.

At the far side of the square, a cord hung between a post and a windowsill. At first Todd thought the line was strung with laundry. But as they approached, he realized that the items were fairly uniform in size, and also color and texture, a sort of leathery brown. Were those sleeves stretched between, with rows of buttons? As they passed, he noticed a bulbous sack hanging from each, and realized with a start that he was looking at a line of octopus. "Calamari on parade," he muttered to Xavier.

"I believe calamari is squid," Xavier said, but he was looking forward and up in their direction of travel. He kept patting his chest, feeling the reassuring presence of his talisman beneath the cloth.

Todd saw a group of Minoans huddled around some type of cauldron. A sizzling sound drifted from it, and so did an aroma. The smell of charring fish burned the back of Todd's throat, and he realized just how famished he was.

He tapped Xavier's shoulder and pointed to the happy group. "Octopus stew."

"*Squid stew* has a better ring to it. You know what's missing here?" Xavier said, his eyes on the Minoans ahead. "Garbage. Refuse. It's a very clean city."

Todd could tell that Xavier was straining to hear words in the conversation of the merchants. The banter continued as the party ascended a stairway. The path led along apartments and other multi-storied buildings, ending at an impressive portico. Along its front rose a colonnade of pillars common in Minoan architecture, their tops wider than their bases. They were painted a startling red, each topped with a black collar.

They passed through the darkened entry. Inside, a rectangular antechamber opened onto three hallways. Each hallway radiated out at odd angles.

"Is it my imagination, or are these corridors crooked?" Xavier asked Todd, wavering a bit.

"Oh, they are. A lot of the rooms we see will probably be irregular, too. The Minoans abhorred—abhor—symmetry."

They stepped through the doorway into an expansive, low-ceilinged room. Xavier hit his head; Todd ducked. The coolness of the inner chamber relaxed him. He realized that he had been gritting his teeth. Xavier was taking in the wordplay, so Todd scanned the great room itself. A forest of columns supported the ceiling, but the room still felt spacious. Light poured in from above, shunted through several light wells. Murals covered the walls. A line of figure-eight shields hung across the longest wall of the room, each covered in multi-colored hides.

Someone gently squeezed his elbow. It was the woman merchant. For the first time, Todd noticed how beautiful she was, with her jet-black ringlets and large brown eyes. Her skin was pale—the patina of a china doll—despite the fact that a merchant like her must have spent a lot of time outside. He wondered if the Minoan women wore some sort of white makeup. They definitely wore lipstick. Her burgundy lips didn't look painted on, as much as they looked brushed gently across the soft skin, whispered color to tint something already perfect.

He realized that he was staring.

With a look of amusement, she guided him across a mosaic on the floor, passing to a low table that stretched the length of the room. She pointed animatedly, using words that were lost upon him. The hulky guy was in the process of seating Xavier already. The people at table—there were a dozen or so—ate Roman style, reclining on pillows. Piles of grapes, squid and an assortment of shelled seafood made up the feast. Large bowls of a sort of porridge made the rounds. Todd watched as the guests plucked food from the center. No plates. No silverware. No fuss.

He noticed that the people around the table, both men and women, seemed to acquiesce to the woman's wishes and follow her lead. She must be a person of influence.

A servant brought in hot flatbread. It was all too much. Todd gratefully sampled everything offered. He had figured out the word or phrase for thank you, or at least some kind of acknowledgement, that many were using around the table. He used it often. The ambience reminded Todd of a feast he had attended with a Persian friend. Hospitality was built into the Middle Eastern culture, as it appeared to be here.

Xavier seemed more reticent to party. He glanced around suspiciously, eating small bits as if he was an actor pretending to be at a meal. But soon, the boss was trying out some of his words, much to the entertainment and appreciation of the crowd. School was in session, and both Xavier and Todd played the part of eager student. Both picked up the rudimentary elements of the language quickly.

Clandestinely, Todd pulled out his bioimager. He shuttered an image of his food, and then of the light well overhead. He scanned the room to make sure he was unobserved. He wasn't: the woman was looking at him and his bioimager. She raised one eyebrow and said several words to him.

"Xavier. Xavier! What's she saying?"

"I think she's thanking you for something. What did you do?"

Todd shrugged. She reached over and gently took the bioimager. She clapped twice. A thin man with disheveled hair came from the back of the room. He wore a smock. The woman handed the man the bioimager. The man bowed to her and turned to Todd, using the same phrase for thank you. What was going on?

Todd sensed that this was a pivotal moment. If he tried to get it back, he might cause an incident. He decided to let the afternoon unfold.

The feast continued, accompanied with the music of several musicians at the back of the hall. The sound of flutes and strings was broken by a hissing. The aproned man was back, carrying a metallic pan with some kind of sizzling meat. Todd immediately recognized the new entrée: it was the bioimager, skinned and done to perfection. The man sliced it thinly and passed around wedges to the attendees. Xavier looked at Todd and grimaced. He shook his head, but Todd had just received his delicacy.

"Bottoms up, Xavier." He chewed it. It tasted like nothing he had ever eaten and like nothing he ever wanted to eat again. Todd leaned over to Xavier and whispered, "Do you still have yours?"

"Yes, and I'm definitely not going to eat it."

The woman and the others seemed very appreciative of what they assumed was the gift. Soon, conversation and laughter again filled the hall. Just as the two travelers were beginning to relax, a clamor issued from beyond a large door at the back of the room. Two huge guards wearing feathered headdresses stepped into the chamber, each carrying very large, very nasty-looking double-headed axes. Their eyes were fixed on Xavier and Todd.

33

People Plotting Plans

A lot of life's little pleasantries made Rex Berringer sweat: heights, peanut butter, his ex-father-in-law. Dresden, the dark Texan, was another one of the biggies. He hated the way Dresden could freeze your blood, even from a distance. With his cadaverous complexion, everything about the man was foreboding.

"Now look," Rex said, keeping his voice steadier than he felt. "I've got it all planned out. Since your crack techies couldn't break through the firewalls or the real walls, I'll go inside and download the files directly to a nice little drive I just got. It's compact and flat and easy to hide."

Dresden stood in silence. Finally, he said, "And how will you get in?"

"I'm a master of disguise. I have something planned."

"Well, my dear P.I., I do hope those plans have a short timeline. And I hope they are more successful than your previous ones."

"I need some time to carry this out. I have to get a few specific cables. I have to rent a truck that can't be traced. These things…"

"Take time, I know. My overlord wants something in hand, in his own personal computer system, within a week. Two at most."

"Getting into a place like ChronoCorp is more difficult than we thought. But there's an event coming up that I can take advantage of."

Surprisingly, Dresden agreed. "Two weeks, then. That may be the end of your employment. Let's hope it's a happy ending." He was so good at sounding ominous.

"Don't worry," Rex said.

"I'm not the one who should be worrying."

M. Carroll, *Plato's Labyrinth*, Science and Fiction,
https://doi.org/10.1007/978-3-030-91709-8_33

*   *   *

Gazing out the window into a spectacular Colorado sunset, the woman rubbed the back of her neck. It had been a long day with new things to learn, piles of paperwork, and some degree of physical exertion. She wasn't used to days like this.

She didn't like playing the role of the mole, but the money was good, and so far, there just hadn't been much to do. ChronoCorp's trip to France had given her new insights, something her "bosses" would undoubtedly appreciate.

34

The Departure

Xavier looked like he was about to panic and bolt. Sweat stood out on his forehead. His face flushed. Todd leaned over and put his hand on Xavier's shoulder. "They're golden ceremonial axes. Religious symbols. The blades are probably as sharp as a kid's plastic sword."

"I see," he rasped.

Todd frowned. It wouldn't do to let Xavier have a meltdown in front of their generous hosts. The weapon-bearers were placing the axes on a dais among a display of garlands. Todd and Xavier thanked their hosts as best they could with their limited linguistic skills, then stepped to the front portico. Todd turned and gave a short bow to the woman who seemed to be in charge, as he had seen several of the Minoans do. The last thing they needed was someone seeing them off at the docks. They needed a clean break. As they left, Todd heard the woman call after them a few phrases, ending with "Shaveeur. Tood!"

"They've learned our names, sort of," Xavier panted. "And a few English words. I learned a lot, too. Let's vamoose."

Todd locked eyes with the woman. Who was she? Why did she have such influence? Why did he care so much?

Crossing the square again, Todd tried to take it all in: the aroma of spices he had never smelled, the feel of the pumice-hewn pavement, the sounds of a language long lost. Some of the tables held piles of ornaments and cookware, perhaps for future sale or barter. Children chased each other across the cobblestones and onto the paths and stairways leading into the mountains.

© The Author(s), under exclusive license to Springer Nature Switzerland AG 2021
M. Carroll, *Plato's Labyrinth*, Science and Fiction,
https://doi.org/10.1007/978-3-030-91709-8_34

Once they had cleared the edge of town, Xavier said, "I learned several names. That could come in handy for our future ventures. But I'm not sure I can remember them."

"We could write them down."

"On what? Nothing lasts."

"You could write them in the dirt and then…" Todd pointed at the folds of Xavier's clothing.

Xavier snapped his fingers. "Of course!" He brought out the bioimager.

"Better do it now," Todd said, glancing behind them. "Let's go into that little grove. What are those, olives?"

"Try one," Xavier said, stepping into the shade of the trees. He broke a twig from a shrub and began to write. "The woman's name was Kitane or Katanie."

"She seems important," Todd said around a fresh olive. "I don't even like olives, and this thing is scrumptious!"

"The big guy, the one we first met, was Duripi. And the tall skinny guy. Raza or Rusa or something. Might have been Araja, too. I couldn't tell where the word before left off and the name started."

"Write it down anyway and I'll get a shot." Todd repeated a name. "Kitane. Kitane."

Xavier wrote the names phonetically, along with a few key phrases. Todd shuttered the platypus polaroid at the ground. It squished, just as it should. But Todd noticed a rough patch of skin on its side. He turned it in the sunlight. Grey splotches covered the "belly." As he turned it back, a few hairs fell out.

"I think our bioimager is sick."

Dread flashed across Xavier's face. Again. Xavier had been panicking a lot lately. "We've got to get it back and download it before we lose any images. Emil and his team are growing two more, but I don't know how all that's going."

A twig snapped behind them. They spun around. On the other side of the grove, an ibex chewed some bark on one of the trees.

Xavier let out a sigh, and the two hurried down the slope. Todd took one last, longing look at the emerald waters of the Aegean. He wished there was time for a swim. He gazed back up the hill, craving just a few more minutes with his new, very ancient friends.

They reached the ridge and scrambled down the side. The ghostly outline of the two pods remained in their places. They triggered their talismans and the pods solidified. Xavier climbed into his and yanked the talisman from beneath his Minoan shirt. "Here goes," he said cheerfully, slamming the hatch shut.

Todd turned to his pod. Closing the hatch, Todd reclined into his couch, gently cradling the bioimager in his arms. As he pressed the tab on his talisman device, he glimpsed the face of Kitane peering over the ridge. *So much for a clean exit.* He shoved his hand against the glass, wishing he could reach out to her. How did she look at the world? At him? What did she believe about eternity or friends or the flowers that grew here?

A flash. And darkness.

* * *

Katya tried not to pace. Pacing in the small Transport Arena was nearly impossible with all the techs crammed in at their stations. The nearly instantaneous return so typical of the time pods wasn't happening. She could feel the tension in her shoulders, her neck. Finally, the alarm rang on the monitor. The subtle glow of the pods heralding a traveler's return was exaggerated, perhaps by the long trip into deeper time. The pods let out a blinding flash. Todd stepped from his in foreign clothing. Katya recognized the Minoan pattern along the hem of his kilt. He sat on the edge of the pod, rubbing his eyes and smiling.

"Well, that was a trip!"

"And you got memorabilia," Katya said. "Retro-fashion."

Immediately, the techs surrounded him, slapping him on the back and giving him high fives. As the excitement died down, they shifted their attention to Xavier's pod. The canopy had swung open slowly, but he had not stirred. Finally, he sat up, swung his legs out and wavered. Kat rushed over and put her arm around his shoulders. "Just sit here a minute, okay?"

"I'm fine," he croaked. He tried to stand, but abruptly sat down again. "Can I get some water?"

Brianne was one step ahead, handing a cup to both Xavier and Todd.

Soon, Kat deposited Xavier onto the couch in the lounge. Todd sat next to a table beside the couch, looking at Kat, Brianne and Emil. "I think the deeper we go, the harder the recovery. Both of us were unsteady on Thera, and coming back was a bear."

"Yeah," Xavier said, clutching a compress to his forehead.

"I wonder what would happen if we took some pain reliever before we left, so it was in our system."

"We would need to run experiments closer to the present," Emil said, "make sure the drugs don't have adverse effects."

Kat let out a long breath. "The pain is a challenge, but if we're going to be sharp during our short visitation windows, we need to be clear of painkillers."

Todd rubbed the back of his neck. "Easy to say here in the comfort of all this paradise." He looked around at the crowded room and laughed.

"No," Katya said. "Emil is right. No drugs until we're home again. Then you can both celebrate with an uber-acetaminophen cocktail. You going to be okay here while we check the equipment?"

Xavier nodded, then flinched.

"Todd, are you up to going with Emil to see to our bioimager?"

"Sure thing."

"Everybody else needs to confirm your data acquisition from the trip and make sure the pods are intact and on line. Talismans, please. Xavier, want the lights off?" He nodded. She held out her hand to Xavier and Todd, grabbed their talismans, and flicked the lights to dim. The others left the room, but as she turned toward the door, Xavier grabbed her hand gently.

"What can I get you?" Katya asked.

"Kat, remember what I mentioned on the plane? What are the chances we might see ourselves, slightly offset in time?"

The hair stood on her arm. She thought for a moment. Somehow, in the dark, it was easier to guess at such a difficult question. "Parallel universe, marginally shifted toward past or future? I'm still not completely convinced that we see ripples in the cosmic background radiation."

Xavier mumbled. "Those supposed fingerprints from an outside universe? Bumping up against ours?"

"I don't think the math is there."

"But what about moving just a few minutes, or even seconds, off of the phase we're in now? What if we travel only moments away and turn to look back?"

"I suppose if two events are close enough in time, the two universes might overlap, may connect. Perhaps there would be some contact. Exchange of information."

"A glimpse into a second reality." Xavier rolled over with his back to her. Katya left him alone to sleep and recover.

* * *

Xavier groaned in the darkness. He listened to the humming of the refrigerator, undoubtedly filled with long-forgotten lunches molding in its depths.

He could hear the gurgle of the coffee dispenser. And he could hear the whispers. They were too faint to make out at first, but like the table talk at the ancient Minoan feast, he began to make out words.

Don't even know if this is working.

Lying down on the job.

Can you hear me?

Hey you. Hey Xavier!

He opened his eyes to darkness and an empty room, but the voice was still there.

You idiot. I told you not to go back. What are you thinking? There's too much at stake here. We need to rein it in.

Another voice. "Here. Just lie still for a while." It was Katya, covering him with a blanket. "I'm sure nobody wants to see more of you than we have to." She lay a set of his clothing next to him and patted him on the shoulder. He could feel his tunic giving way beneath the blanket. Soon, the Minoan garb would be dust. In the ageless excursion of time, soon they would all be dust.

She closed the door. The voice was gone.

35

Reporting Back

The bioimager was dying. There was no way around it. Emil held it up in the light. The surface glistened, and the lobes on the side pulsed weakly. The pelt of fur lay flat against its side. On its belly, gray-green patches spread in a bilious stripe across the naked flesh. Electrodes led from several orifices to an electronic/organic hub, then to an HDMI port.

"What's with this thing?" Todd asked, placing it back on a metal pan. "Can we give it vitamins or penicillin or something?"

"I'm beginning to think our bioimagers may have a lifespan. Their operational lifetimes might be shortened by the number of Q-slips we take them on." Emil seemed overly nonchalant. "I'm sure it doesn't help to roll them in a grassy French field."

"If this thing dies, what's our plan B? Is there one?"

Emil smiled as software transformed a raw image of gray snow into a color shot of a Minoan portico. The brick-red colonnade fronted a mural of royal blue background with white lilies. Emil looked up from the monitor. "We've got them. Good." He sat back in his chair and put his feet on his desk. "Plan B is the next generation of bioimagers. We've been growing them for a month, and two are nearly mature."

"How soon can we use them?"

"We'll test the two most developed ones next week. If they check out, we should be ready to go a week from Friday." He did a double take and leaned toward the monitor again. The screen displayed a Minoan ship, its cargo resting on a stony jetty. "Wow, the things you guys got to see."

Todd nodded. "Yep. Pretty amazing."

© The Author(s), under exclusive license to Springer Nature Switzerland AG 2021
M. Carroll, *Plato's Labyrinth*, Science and Fiction,
https://doi.org/10.1007/978-3-030-91709-8_35

*   *   *

"I tell you, the room was flooded to bursting with technicians and engineers. They all huddled around the main monitor; Emil had to mirror the images on several other screens to let everyone see." Katya tried to contain the excitement in her voice. She wanted to come across as strictly professional. Detached. It probably wasn't working. Even Wellington could barely sit still on her lap.

"And you could see details on the pithoi?" Mila van Dijk asked breathlessly, her wrinkled face accentuated by the harsh light of her own computer monitor. She rocked in and out of the computer camera's field of view, too excited to sit still.

"Oh, yes, sculpted frills and grape leaves, and they were all painted. Beautiful things."

"Well then," van Dijk said in a satisfied tone.

"So I've asked Xavier to let you come in for a glorified slide show of what they saw. It's the perfect excuse for you to come in for a 'meeting.'"

"Yes, that will do nicely. Thank you, Katya. You just might make a matchmaker yet." Then van Dijk's attention drifted. She seemed lost in thought as she muttered, "Q-slip number one was a success, and trip number two is imminent."

"That's the long and short of it," Katya said. "So about those images?"

"I'd love to come see them. At ChronoCorp. And I'd love to see how you do it. Using a laser to break the quantum barrier? Intriguing."

"A big corkscrew of three lasers, actually. We can show you that, too."

"How soon can we do this?"

"Next week? Things are moving fast now."

"Thank you, Katya. And remember, not a word to Xavier. I don't want him getting spooked, thinking I'm checking up on him or something. I just want to join in the excitement of exploration."

And the excitement of romance, Katya thought.

36

Dinner with Bradley

Katya had been firm about dining in Denver rather than "the Fort." Of course it was easier for Brad; it saved him a trip. At the same time, it was a shame. Brad was hoping for his tour of ChronoCorp. Apparently, for whatever reason, they weren't ready to admit him into the Holy of Holies.

He dug his knuckles into his eyes and yawned. It had been a long day of lectures and poster sessions. All day he had dreaded this dinner. Katya Joshi seemed a schizophrenic contradiction. Her dark hair framed a face that sometimes favored her Indian background, and at others revealed her Slavic side. It was an elegant combination, he thought. She was a disciplined physicist and, according to his web searches, a rigorous researcher who got verifiable results in inventive ways. When he watched her with her coworkers, she teased them mercilessly, dishing out as good as she got.

Still, he got the feeling she was hiding something. With her father, you got what you saw. Ajit Joshi was an open book. Brad's work had brought him alongside many paleontologists who held their dig sites—and insights—close to the vest, but Ajit talked about his work and personal life as if they belonged to everyone around him. Brad was all for confidentiality, but Katya's seemed to spill over from professional to personal. He always felt as though she was generating a force field, her own little magnetosphere, deflecting his efforts to break through and warm her up. Not that he had any burning desire to do so. She was more than vaguely interesting, but it was ChronoCorp that drew him. He wanted to see its strangely clandestine workings, and she was his in.

Still, their last meeting left a sting. He had said or done something to offend, and he owed her dinner, at least. He really didn't want to screw up this

© The Author(s), under exclusive license to Springer Nature Switzerland AG 2021
M. Carroll, *Plato's Labyrinth*, Science and Fiction,
https://doi.org/10.1007/978-3-030-91709-8_36

night. Besides, if he did, Ajit would probably hear about it, and he still hoped to collaborate with the man on a project in the Gobi Desert.

The hostess at the entrance to the hotel's restaurant was talking animatedly to someone. She nodded, swiveled, and pointed toward Brad's table. Katya came through the door and into view, and headed over. He stood and pulled a chair out for her. There was an awkward instant when both of them debated what kind of greeting to give each other, until they simply shook hands.

"Well, well," she said. "Quite the gentleman."

"Chivalry may be on life support, but it's not dead."

She dropped her napkin into her lap. "I, for one, am glad."

"Are you a bit old-fashioned?" he asked. It hadn't come out the way he wanted, but the comment was beyond fixing.

"I'm not sure there's anything old-fashioned about being gallant."

"Yep," he jumped in. "I completely agree." Why was he sweating?

"Of course, it comes in different flavors, I suppose. Servanthood? Self-sacrifice?"

A man two tables away was raising his voice. Finally, he reached a crescendo, stood, and threw his napkin on the table. "That's it!" he bellowed, then left. The woman at the table looked down at her spaghetti.

Brad leaned toward Katya. "We could add to the list a moderated volume."

"And no napkin tossing."

They ordered wine. They agreed about what a shame it was that Katya's father was in the field and unable to make the conference, let alone this dinner. They talked about small things and ramped up to the bigger ones. Maybe the wine helped, or maybe Brad was just relaxing. He had been out of the dating scene for a while.

He couldn't get the couple at the next table out of his mind. Why had they fought? Was the man just a big oaf, or did he have a valid gripe? He had left so dramatically. Was she going to be okay? Should he do something?

"I was hoping you and I could chat about some things, about your work and father and ideas on the Big Bang and such, but…" He shrugged and looked over at the woman in distress. "I suppose we could invite her over to join us." He got no reaction. "Only if you think it's a good idea."

"See, there you go, being self-sacrificing. Yes, I think that's a very sweet idea."

The woman stood and left abruptly.

"Theoretically," Katya added.

"There she goes," he said, watching the woman make her exit.

"Yes, but you still get extra credit." Katya cocked her head and offered a gentle smile. "I'll bet you're great with kittens and puppies."

There was that tone again, the tease. Was she toying with him? They were doing an uncomfortable dance around each other, and he couldn't tell who was leading. He had once read that Pluto and its moon Charon circled a point somewhere in between them, instead of Charon going around Pluto. Maybe they were like that, circling something in the center between them.

"Kittens and puppies? I'm allergic. Mostly. But I did buy a loaf of banana bread at the local Dumb Friends League benefit."

"That counts." She studied the menu. "So, the conference. Dad was very sorry to miss it. How has it been?"

"Exhausting. I heard a keynote this morning from one of my heroes, a Canadian paleontologist who's the world's leading expert on horned dinosaurs."

She peered over the top of the menu. "Like, ah, Triceratops? Styracosaurus? Pentaceratops?"

He flinched inwardly, feeling as if he had just talked down to her. "I forget what kind of childhood you must have had with Ajit and your mom at the helm. Yes, the ceratopsians as a group. And then there was a panel on possible acoustics of the lambeosaurs. I didn't have time for lunch. Even this menu is starting to look good."

"I'm sure it's high in fiber."

The waitress appeared. Katya pointed to Brad and said, "Can he just have one of everything?"

The woman was young and probably new to the food services. A look of confusion spread across her face. When Brad and Katya laughed, she tried to join in, a little late.

"You have a nice laugh," Brad told Katya. He sensed that he had made another misstep. By way of explanation, as if it would help, he added, "It lights up your face." And that probably made it worse. "Look, Katya, I felt I owed you at least a dinner. I did something that got us off on the wrong foot back at Princeton, and I wanted to make it up to you."

"You want me to take pity on you?" It was an unkind dig. But suddenly she was grinning. "Brad, you have nothing to apologize for. I think we may have different perspectives on a few things, that's all. Nothing to get in the way of a nice dinner, right?"

She was letting him off the hook. Maybe it was because the waitress had shown up again, but he felt as though he wanted more than just getting through this dinner. He wanted to make things right. He forged on, ordering from the menu. "Sorry I'm waffling," he told the waitress.

"Waffles? We stop serving breakfast at eleven. Sorry."

Katya hid her face behind her menu. Brad looked away, stifling a grin.

Katya ordered, handed her menu over, and looked away while the waitress left. When she looked back at Brad, he lost control. Laughter paralyzed both of them.

When they had recovered, Katya became serious. "My dad says you're not a desk jockey. That was unfair of me."

"No worries," he said, not waiting for a formal apology. It was his turn to be gracious.

"He said you've been on digs all over the world. So, globetrotter, what was your favorite place?"

"Vega Island, off the coast of Antarctica. Tough to get to. We had mostly bad weather, which in Antarctica is saying something. Even good weather is dicey."

"And when it wasn't blizzarding, what did you find?"

He noticed that light in her face again. It played across her cheeks and danced in the blue of her eyes, the same blue he had seen in the glacial floes of Antarctica. But there was nothing cold about it. "No kittens and puppies, I can tell you. Aside from penguins and those nasty Skuas—"

"Nasty Skuas?"

"Dirty brown birds. Some liken them to seagulls, but I think of them more as rats with wings. But besides all that, we found non-avian dinosaurs. Beautiful ones. That's what my paper was on yesterday afternoon. A new hadrosaur we described. People have found teeth there before, but we found a skull complete with a beautiful crest. And quite a few vertebrae."

Brad tried to hit a happy medium as he described the fossil finds. She was no paleontologist, but she was a scientist, and she was daughter to one of the best dinosaur hunters around. Thankfully, the bread came. Brad turned the conversation to her.

"So what's a world-class physicist doing in a place like Fort Collins?"

"World class?" He heard skepticism in her voice. "Someone's been telling stories."

"You can guess who. He did mention a few awards you have on your office wall."

"Oh, no, I keep all my accolades on my refrigerator, next to my crayon diagrams of atomic structures."

"I keep mine under several layers of dust in the basement."

"Like any good paleontologist should."

Brad looked at her, and she didn't take her eyes from his. He let the silence linger just long enough for her to realize that he had noticed her putting off an answer. Again.

He said, "I understand the need for professional confidentiality. What I don't get is this: there seems to be a tie between my hobby—Waterhouse Hawkins—and your work. Quantum physics seems to be pretty far from outdated dinosaur art."

"Quantum physics is the language of the very small. The language of time and space. So, now comes the speech."

"Shall I get you a white board or do you have Powerpoint?"

"I'll just wave my hands a lot." She leaned forward and dropped her voice. Her purple blouse shimmered beneath her charcoal business jacket, a brilliant V of honey-brown décolletage marking the center. Brad made a conscious effort to not let his eyes wander down her low-cut collar, tasteful as it might have been.

"I know you suspected something was going on when a gaggle of physicists came plowing through Princeton to see a bunch of dinosaur drawings."

"The thought had crossed my mind."

"And just what did you suspect?"

He shrugged self-consciously, glanced back and forth, and took a sip of wine to delay. He set his wineglass down and decided to be frank about his suspicions. "Did you send someone through time to the Mesozoic?"

Katya grinned. Her smile was disarming. "Nothing so exotic. But our work does dovetail with yours."

Brad shivered. He couldn't tell from her expression, but her decibel level—almost a whisper—spoke volumes. Katya slid a piece of paper across the table. It looked uncomfortably official.

"What's this?"

"Confidentiality agreement. We want you to do some consulting for us."

This little bombshell took their dinner date right off the frame of romance and onto the professional track. Brad was surprised at his own reaction: disappointment. But he was also captivated. He read the short contract and signed.

Katya folded the paper and slipped it into her purse. "Right now I can tell you that we need you to be a sort of monitor for us. Remember when you commented on Hawkins' rendering of the Iguanodon? The posture?"

Brad remembered the hair standing up on the back of his neck when he first saw the image. He remembered forcing the quaver from his voice, the trembling from his hands. "Tail flat. Streamlined stature. Weight on the back legs. I'll never forget it."

"That's exactly the kind of thing we're looking for. You can be our monitor. We need you to look for elements in the Hawkins drawings that don't belong. Things that seem out of place."

"Out of time, you mean?"

"See? You're not just a pretty face. And when you have your tour, we'll be able to tell you more."

This was his chance, and he wasn't going to let it slip by. "And when might that be?"

She smiled, swirled her wine in the glass, and took a sip. "We have to reach a research milestone first, but once we do, Xavier will be happy to have you come. I think it will be soon. I hope. More wine?"

She poured. Brad knew that was the end of the discussion. For now.

He enjoyed the banter of the evening, and when the bill finally came, he was feeling the warmth of the wine. The waitress placed the tray on the table between them, not committing to either side.

"You've been a patient man," Katya said. "We'll have you up to the lab soon." Brad's hand rested on the table next to his plate. Katya reached over and put hers firmly on top. "But it can be a dangerous place. You will need to follow me closely, every step of the way." Her tone was melodramatic.

Brad looked down at their hands, turned his upward so that Katya's was in his palm, and raised it nearly to his lips, but didn't quite make contact. "Yes, my liege."

She pulled a Mona Lisa smile. "You know what Edith Wharton told Henry James, don't you?"

He gently released her hand. "I'm not up on my literature of the Gilded Age."

"The flavor starts at the elbow."

He sat back in his chair, trying to keep a debonair affect. "Well, we'll just have to see about that. One of these days."

Katya said, "I did bring up business, so this is a business expense." She reached toward the bill.

Brad couldn't let that happen. Suddenly, he realized that he wanted this to be more social than business, no matter how old fashioned. He lunged for the tray. His elbow brushed against his wine. The glass splattered crimson across Katya's business suit. He was reliving the nightmare of the last time he saw her at a conference.

"Old habits die hard," Katya said.

"I'm an idiot," Brad huffed, feeling the flush of wine settle on his cheeks.

"No, you're just a kittens-and-puppies kind of guy. I appreciate the effort. And I will let you pay the bill. It was a lovely dinner."

Her comment was magnanimous, but her expression said, "Not again," or even worse, "This guy *is* an idiot." As she ran off to the restroom to salvage her business suit, he tried to estimate whether this star-crossed relationship was worth recouping.

37

Sleep is for the Weak

Up to now, a voyage to ancient Thera seemed amorphous, abstract. But now that Katya was preparing to go, it had a sharpness to it, and she liked it. Life was like that. You could live it within dull, predictable borders, or live it on the blade-edge of excitement, of *the new,* with a spot of calculated risk thrown in. She preferred it this way. She had learned from her parents.

Still, she had put off the next voyage until they made sure everything worked as it should. At Xavier's insistence, the techs had worked around the clock. The pod systems had behaved as predicted, and the second bioimager had matured to readiness. The only facet of the team that was not back to normal was Xavier.

"Boss, just let us do our job and you take a break," Todd pleaded. "For just a little longer."

"Sleep is for the weak," Xavier chided. "Besides, someone needs to supervise Katya. You know how she gets."

"And just *how* do I get?" Katya asked.

"Creative," Xavier said.

"Let me go," Todd said. "You have to get that paper done and submitted so we can do our press conference thing."

"And so you can present at San Francisco," Kat added.

Xavier buried his head in a port at the side of the pod. He began to sing, quietly. "I have never met Napoleon, but I plan to find the time…" He spoke over his shoulder, "Todd, we need to get our scholarly ducks in order. Otherwise, this is all for nothing. Somebody else might be coming on our heels. You know how research is."

© The Author(s), under exclusive license to Springer Nature Switzerland AG 2021
M. Carroll, *Plato's Labyrinth*, Science and Fiction,
https://doi.org/10.1007/978-3-030-91709-8_37

Katya stepped over to the pod Xavier was examining. She crossed her arms, rested her elbows against the rim, and looked him in the eye. "Xavier, there's some wisdom in what Todd is saying. We've worked out the kinks. Now it's time to bring others in. You don't need to be there at every turn."

"It's not that I don't trust my wonderful team, Kat. But I want to be there, on Thera, one more time." She saw an intensity in his eyes, the kind of wild energy more fitting of a religious zealot. "Then we'll start several independent teams. Besides, I know the language. Sort of." He shuddered, swiped beads of sweat from his forehead, and turned back to studying the pod.

Katya backed away toward Todd. Todd mouthed the word, "See?" Then he ushered Katya down the hall and into his workspace. "Kat, I'm worried about you."

"Gee, Todd, I didn't know you cared. Besides, it's not me you should be worried about."

"That's just the thing. The boss has had too much travel. It's knocked something loose up there." He tapped his forehead. "There's just something not right about him, and it's more than exhaustion. Look, Kat: when you're out there, it's just you and him. And there's no AAA in ancient Thera. I checked. I'd think twice before running off with him to some distant place like that."

She was nodding her head yes, but she knew on a visceral level that it was time for her to go to the heart of the Minoan Empire, and if she had to go with Xavier, so be it. She glanced at Todd's computer. His desktop displayed the famous Minoan painting of the woman known as *Le Parisienne*, a lovely face with porcelain skin framed by black curls. Katya nodded at the screen. "That's nice."

"She…reminds me of someone."

His tone was strange, she thought.

38

Inside Information

The Ambassador's secretary put the call through immediately, per his instructions. If that bumbling P.I. wasn't going to produce, at least the Ambassador's mole would.

"And what news do you have for us?" he asked. "Progress at ChronoCorp?"

"Indeed, Lou. Their next trip should take place within the next two weeks, and my source thinks it will be considerably sooner."

"Same destination?"

"Yes. The Minoan capital."

"Timeline?"

"Thirty-five centuries ago, give or take."

"So," the Ambassador said professorially. "They've already made the trip once. All that way back there. And they've made at least two other tests we know of. If they're successful again, they've got all their problems ironed out—the same problems we've been dealing with."

"Correct," the mole said.

"I suppose that is not your concern, for now. Mr. Dresden and our private investigator have come up with a strategy of sorts. But you must be standing by to step up your involvement. Up to now, your reports have lacked a sense of…" he searched for the right word, "technical details? Particulars? Even minutiae? We need minutiae."

The mole paused.

"Will that be a problem?"

"No, of course not."

© The Author(s), under exclusive license to Springer Nature Switzerland AG 2021
M. Carroll, *Plato's Labyrinth*, Science and Fiction,
https://doi.org/10.1007/978-3-030-91709-8_38

"That is fine. I look forward to more progress in your next report. I'd hate to see certain details of your financial past leaked to the public. They could be misinterpreted. Clear?"

"Clear."

The Ambassador hung up. He thought he had been finished with that woman. But when the opportunity showed itself at ChronoCorp, he had to bring her into it, of course. *Sad how the past continues to haunt us*, he mused. Just when he thought he had liberated himself from her talons, circumstances brought them together again.

He stepped to the window, rubbed his eyes and flicked a switch on his phone set.

"Sir?" came the response.

"Get me a driver."

"Yes, sir."

The Ambassador walked out of his office. He had to think, and it would be easier outside of town, up the Poudre River at the mountain facility. Fresh air, surging mountain waters, and rows of cafeteria tables awaiting hordes of ancient Romans. These were the things that lifted his spirits.

*   *   *

Mila appeared at ChronoCorp's lobby promptly at ten, just as agreed. Katya and Todd met her. Todd noticed a subtle tilt of Katya's head, something in her eye, an acknowledgement passing between her and Mila van Dijk.

Katya held out her hand. "Ms. van Dijk."

"Dr. Joshi. Dr. Tanaka."

"Welcome," Todd said cheerfully, shaking her hand.

Katya said, "So glad you could make it, Ms. van Dijk. Let us show you to the conference room. Just this way."

Todd wondered what Mila van Dijk would think of their humble meeting room, especially after coming from a mansion where some of the bathrooms were larger. But Mila seemed comfortable as she sat in one of the chairs at the long table. Xavier stepped through the door, reminding Todd more of Quasi Modo than the head of a research team. He shook hands with Mila and sat heavily by a control panel recessed into the table. Todd watched her. She couldn't take her eyes off Xavier.

"Well, gang, I guess it's show time," Xavier said. "Ms. van Dijk, with your permission?"

"Please, call me Mila. Go on. I'm all ears."

Xavier triggered the projector, the room lights dimmed, and Xavier sat back. "Todd and I travelled together. Todd will do the honors. Todd?"

"Sure," he said, eyeing van Dijk. He thought he could see her mouth watering.

"We arrived in the shelter of an olive grove, just adjacent to an earthen berm. It was a nice, secluded spot for the pods."

"Pods?"

"We travel in individual pods," Katya interjected. "We'll show you. They're a bit like rounded coffins with clear lids."

"I see." Mila studied the blank screen.

Todd hit the keyboard. A dark, poorly focused image of a wall came into view, with the blue ocean behind. "This was the first shot we got. We're downhill from the pods, in a warehouse area adjacent to a port."

Mila shifted in her seat. This was the reaction Todd had anticipated: expectation mixed with disappointment.

"I was quite dissatisfied with this first image, too." The screen faded to a glorious view of the harbor, rich azure and emerald waters lapping against a black sand shoreline. Uphill, beyond a curve in the beach, a scattering of red and tan buildings jutted from a green landscape. Todd could hear Mila's intake of breath.

"That's…this is spectacular," she gasped.

"It was," Xavier said quietly.

The next few images revealed detailed views of the various pithoi, with their painted and sculpted elements of vines, grain, and Minoan design patterns. Then came one of the buildings, a warehouse standing near the lapping Aegean waters. Todd explained a gap in images where their Minoan friends had clothed them and led them to the nearby village.

The next images built to a crescendo, with Todd reveling in his tour-guide part. "And finally," he said theatrically, "it was time to go." He revealed a last image, the entryway to the building complex where they had dined. Tapered Minoan columns framed a dark doorway, with the typical double-headed axes mounted on the lintel above.

The screen faded to black. As the lights came up, Mila van Dijk clapped. "Stunning. Absolutely stunning."

"Do you think the press will like them?"

Mila leaned forward. "The press is going to go ape-shit." The room went silent. Mila said, "Oh, excuse me. But they will."

Laughter erupted.

Xavier said, "We'll need to exercise some control. We need to have our first release timed to our paper's acceptance. Then we'll save a few images to debut at the conference." Todd folded his hands on the table. "We'll be presenting two different subjects: what the popular press will undoubtedly call 'time travel,' and the details of the Minoan architecture we saw. There will inevitably be tabloid journalists there. At the other end of the spectrum, we need to be prepared for some very technical questions from archaeologists and physicists alike."

"And I would hope I would be on the list of invitees?" Mila cooed.

"You'll be at the top," Xavier said. "Because, as you well know, we could not have done this without you."

Mila smiled. Todd thought she was avoiding eye contact with Katya. *Curious*. Todd looked to Xavier, who had gotten distracted and was looking away. But Mila van Dijk had him pinned with her eyes. She wasn't studying him. She was adoring him.

Todd led Mila's brief tour. She was treated to views of the lab with its bioimagers (she likened them to beef heart sous vide), the Transport Arena (she suggested wallpaper), and the room adjoining the laser tower.

Mila crossed her arms as they walked, nodding and mumbling, "Very impressive. Exceptional." Once Todd had led them back to the lobby, Xavier rejoined the group.

"Well?" he asked her, bouncing on his tiptoes with his hands in the small of his back.

"Impressive. Xavier, may we talk in your office? I have some figures to go over with you."

Todd fought the urge to say something about Mila going over Xavier's figure.

To the others, Mila said, "I can't tell you how much I appreciate this. This may call for another celebration dinner at the house."

Todd, Xavier, and Katya all asserted their agreement. Todd wondered what kind of ice sculpture she would have next time around.

* * *

Emil placed the bioimager 2.0 in its cradle. This one had improvements in operation, but he hoped it would also be longer-lived than the other specimens. Several had died before making any trip, while others gave up the ghost almost immediately after making the Q-slip. This one promised longevity, thanks to a sequence of DNA his colleagues had spliced in at what they hoped was just the right place.

As he attached electrodes to several orifices, he heard someone step into the lab.

"Hi, Emil. What are you up to?"

It was the new recruit, Brianne. He could tell she was a quick study and a sharp cookie. People who weren't sharp didn't get past Xavier's hiring process. One of the other techs had told him that she was "a babe," but Emil hadn't noticed. His work kept him too busy for such things.

"New bioimager," he said, not looking up. "I'm just testing it to see if it's healthy and ready for the next Q-slip."

Out of the corner of his eye, he saw her shiver. Most people seemed to find his bioimagers revolting. Emil thought of them more as pets with not quite enough skin.

"Yes," Brianne removed her glasses and pointed them at him. "About that. The target is the Minoan capitol again, right?"

"Correct."

"Is that the farthest anyone has gone?"

"Far? Don't forget: in this place we talk about position in both space and time. This is the farthest anyone has gone in both dimensions, yes."

She had made her way to his workbench. Now she was standing next to him, looking over his shoulder, brushing up against his side. The tech had been right. Maybe Emil should pay more attention to his workmates.

"Do you think they'll let people like you and me go?"

"Oh, sure, eventually. When it's safe. Q-slips are still firmly in the realm of *experimental* right now."

Brianne waited for a few moments, letting Emil work. Then she said, "Those lasers. Behind the building. They must take a lot of power."

Emil turned from his bioimager and looked at Brianne. "Oh, they do, and it's been a problem. Under the building is essentially a giant capacitor that stores energy. When it's time for the quantum vortex—the Q-slip—we discharge a huge power surge into the triple laser system, and the charge has to go through all of them at once or you get delayed frame dragging from one to another."

"Three lasers?" Brianne prodded. "I thought there was one big spiral."

"Three spirals. Three lasers in tandem, corkscrewing up the tower. We tried one but it just wasn't powerful enough."

She nodded, pinching her chin, suddenly deep in thought. "Makes sense. Three. Yes, makes a lot of sense." She turned toward the door. "Gotta go, Emil. Good luck!"

"Thanks, Brianne." He followed her out the door with his eyes, and then refocused on his bioimager. It pulsed and undulated. For the first time, he shivered.

39

Deviations

Detours. Life was full of them, Brad reflected. Some were trivial; others could be earth-shaking. There were simple detours like road construction or changing your mind about where to get takeout. Then there was the global pandemic that cancelled his first major presentation at a world congress of paleontologists. The death of his father just before his cousin's wedding. Most of the deviations in his life had been unavoidable and uncontrollable.

But this one was of his own bidding: a detour to Princeton on his way home. He had to examine the other Hawkins drawings.

Both Evie Long and Chad Markham had taken valuable time and effort to provide him access. Evie stationed him at a light table in a sort of clean room, pristine white with vents in floor and ceiling. She asked him to don a baby blue mask, matching glossy shirt and thin gloves.

"Gee, I feel like I should be working in some laboratory," he told Long.

"We're preserving history here," she replied.

Brad sat at the table and squared his shoulders. He leaned in, studying the first drawing. Like the others, the fragile illustration was mounted to a backing piece, hinged to it so that three edges were free. On the outside edge, a framing piece protected the sides. A front board, hinged at the top, bore a label with the collection and the item identification number.

"Just dial six on the wall phone when you're done," Evie called from the door.

She drifted from the room like a specter, leaving him with a portfolio of Waterhouse Hawkins treasures. The first image was a sketch for a painting done for Princeton, a small mural Brad had seen in person. It depicted a group of Hadrosaurs fleeing a toothy predator, one Brad didn't recognize. But

© The Author(s), under exclusive license to Springer Nature Switzerland AG 2021
M. Carroll, *Plato's Labyrinth*, Science and Fiction,
https://doi.org/10.1007/978-3-030-91709-8_39

he recognized the Hadrosaurs. If he was not mistaken, they were based on a skeleton model Hawkins had done while working with Joseph Leidy of the Academy of Natural Sciences in Philadelphia. He based the head—then missing—on the skull of an iguana. The sketch, as with the finished painting, showed snake-necked plesiosaurs and crocodile-like reptiles slithering through a river. To one side, a pterosaur spread dragon wings. Nothing about the drawing rang false. The quaint view of prehistoric life had all the right elements.

He moved on to another, this one displaying a lovely, bristly Hylaeosaurus. Modern paleontologists had far more specimens available to them, and knew that this creature belonged to the clan of armored, tank-like dinosaurs. Its spines likely protruded from its flanks, but Hawkins crafted his beast as a giant lizard with a row of dramatic spikes along its back. Again, no change here.

The next rendering was a preliminary sketch for a piece he had seen at Princeton, an oil painting on a large canvas depicting life forms of the Jurassic. This study was a simple line drawing of three predators lunching on a downed dinosaur. The drawing would become the central area for Hawkins' oil painting. He took another look. There were details here he had never seen before. The creature in the back reared on two legs with a strangely modern stance. It held its tail erect, something not common to drawings of Hawkins' time.

Brad blinked, looking again. Along the spine of the three predators lay rows of fine texture. The patterns didn't look like reptilian scales. Could these be feathers—something that had not shown up in dinosaur art until the last several decades?

He snapped a shot with his cell phone, jotted some notes, and moved on. The next image was a full-color rendering showing a herd of long-necked dinosaurs. But the wandering beasts on the ground—so often shown as lumbering reptiles dragging their tails—showed a spark of energy. These creatures practically danced across the prehistoric plains. Whip-like tails waved through the air and their necks, lowered to nearly horizontal, allowed for a fast track. One of them reared back on two legs, unheard of in Hawkins' time. The image sent a charge of electricity through Brad and raised gooseflesh on is arms.

Several others seemed unchanged, but he had seen enough. Here were the alterations Katya had been looking for. He wasn't certain of their full import, but he guessed they were important to what was going on at ChronoCorp. He had to let her know, but he had to catch his flight first. He would type up his notes as soon as he returned, study the photos more closely, and call for a phone consult with the results. He sent a quick text as he climbed into a taxi for the airport and home.

MISSION ACCOMPLISHED. CHANGES NOTED. JUST WAIT TILL YOU SEE!

Sadly, Katya would not be near her cell phone for more than three thousand years.

* * *

Mila left Xavier's office well after closing time. Katya waited patiently. Finally, when the coast was clear, she went in to his desk. She rubbed her hands together and said, "So…tomorrow's our big day."

"Yes, yes it is." He was clearly distracted.

"And how did it go with Ms. van Dijk?"

"Mm? With Mila?" His cheeks flushed. "Well, you know."

"Know what?" Katya said innocently. Although she had played matchmaker, and the results were technically none of her business, she was hoping for the full scoop. Judging by the glimmer in his eye, she might get it.

Xavier sighed and shook his head with a wan smile. "You just never know what someone is thinking. It's always a surprise. I mean, I'm no Don Juan, but I've had my share of romantic relationships. But this? I didn't see this coming. Maybe I was distracted by the work, or by her official role, or by these fucking headaches I keep getting. But she made her feelings fairly clear right here in this office. Moments ago. I'm not sure what to do with all that."

"Seems like you should be asking yourself how *you* feel about *her*."

He met Katya's eyes. "Yes, that's right. Mila van Dijk is…is…" He dropped his gaze to the table. "She is challenging and independent and exciting. I enjoy her thirst for knowledge. Her questions go deeper than surface curiosity. She wants to know how things tick, whether we're talking Minoan square-rigged ships or triple spiral lasers. I like that. I like that very much."

"Good for you," Katya said. "I wish you guys the very best."

"There's a long way to go before people start wishing each other things like that. I just had this revelation today. I'm processing."

"Process away," Katya said, inwardly congratulating herself on a job well done.

* * *

Xavier felt a sense of accomplishment at the headway he and Todd had already made on their first scientific publication. Todd had taken the lead on the project—a paper about their research into ancient Thera—since Xavier was preparing for his second trip back. Photos from the bioimager would form a pivotal piece, and Emil was working on patents and a research paper about the freakish platypus polaroid at the same time.

These were the things running through Xavier's mind as he lay in the pod, awaiting the Q-slip. He felt better than he had in days, although he couldn't shake the headache, and his left hand seemed to have developed a tremor. Still, they were well on their way to reaching a point where they could reveal their work to the world. That made him very, very happy. He wondered how Katya was feeling. She had been out only twice: once to the office of days past, and once to Central Park of the 1800s. Now, she was embarking on the big trip to Thera. He was happy for her, too.

As he braced himself for the Q-slip, he began to think of the pressure. Yes, their progress had been good. But there was a lot of competition in the sciences: competition for funding, competition for press, competition to be the first (like Edmund Hillary or Yuri Gagarin). He couldn't deny that sense of urgency, goading him on.

Who ran the Primus Imperium and what were they after? Were they, too, investigating the great timeline of the cosmos? ChronoCorp must not only be the first, but also the best. At the same time, he felt the compunction to police the entire field of time travel, to make sure no one was doing this type of work for the wrong reasons. His reasons were pure. Absolutely. Weren't they? The search for origins of the universe, the documentation of past civilizations. That was pure, was it not?

Mila had added yet another layer of complexity to everything on his full plate. And there were the ghosts, the voices that spoke to him in the dark when no one else was at the lab. It made him wonder: could he travel back a few days or weeks and give himself a message?

He closed his eyes and waited for the coming torrent of pain in his head. Perhaps this time would be better than the last.

40

Deja Stew

"Mark Three," Todd called into a headset. "Power up."

"Lasers at full," Brianne called out.

Todd glanced at the shovel on the wall. *Sometimes it just piles up dark and deep.* He was prepared for the flash, for the power surge and the darkness. Kat had given him a preview. As the lights came up again, the pods lay empty. For better or worse, Xavier and Katya were gone. Would the timespan cause a quantum drift too far? If they arrived at the same place but a century later, they would materialize in open ocean off the coast of a smoking, volcanic rubble pile.

He looked across the Arena at Emil. Emil was watching him, too. "They're going to be just fine," Emil smiled. "Better than fine. This trip will give us our ammo for a press conference."

"And a whole bunch of peer-reviewed papers," Brianne added.

Todd stood stiffly, looking at the empty, coffin-like pods. "Yep," he said. "Yep, it will." But he couldn't rid himself of the thought that with each push into the past, they were pushing their luck.

*   *   *

The trip was far more disorienting than her foray to Central Park had been. Todd and Xavier had warned her, but Katya wasn't prepared for the adrenaline rush, the blistering ache at the back of her skull, the glowing black spots at the edge of her vision. Yet she knew it was going to be worth it all.

© The Author(s), under exclusive license to Springer Nature Switzerland AG 2021
M. Carroll, *Plato's Labyrinth*, Science and Fiction,
https://doi.org/10.1007/978-3-030-91709-8_40

She unsealed the lid, sat up, and felt the Mediterranean sun on her face. She took in a deep breath. Salt water. Kelp. Smoke. Spices. Wildflowers. Seagulls circled overhead.

Katya's pod faced a low embankment, sheltered from the uphill direction of the slope. She stepped out and wheeled around. Xavier's pod was just below hers, and beyond it lay the terraced slopes cascading down to the impossibly blue harbor. Grasses and wild saffron softened the rocky slope. Ships bobbed in the waters beyond a series of jetties, perhaps the concentric circles Todd spoke of. She knew these ships were likely headed to or from Crete, as Thera was on the main route between Crete and mainland Greece. As she studied the layout of buildings at the coast, Xavier opened his pod. He stepped out, wavered, put his hand to his forehead and looked up at her, grinning.

"Amazing, isn't it?"

Katya realized that she couldn't speak around the lump in her throat. It was overwhelming. She was witnessing the reality of a world 3,500 years removed from everything she knew. She nodded at him.

At that moment, Katya wanted to dash to the water, tear off her clothes, and revel in the Aegean waters. But then she remembered the tight agenda they had. Step one: find clothes at the warehouses. Step two: reconnoiter the entire complex of buildings at the port, imaging as many as they could without being seen. Step three: head uphill to the courtyard to chart any differences since the last visit, and step four: finish documenting the buildings on the far side. Todd's plan was to assemble a detailed blueprint of the Minoan courtyard, using Xavier and Katya's reconnaissance. His project would lend priceless insight into Minoan city planning and Neopalatial architecture.

Her pod rested against a crumbling wall. Katya considered the ruin, an ancient wall even in this ancient time. Here was a lesson: this time and place was part of a continuum, just as her own time was. This wall hearkened back to an era even more distant than this one, built by real people with individual lives and dreams and aspirations.

The pods had arrived late in the day. Katya guessed sunset was an hour away. The ships appeared to be moored for the night. Cargoes lay stacked along the shore or on the docks, waiting to be loaded. Timber seemed to make up a large percentage: fir, cedar, oak, and cypress, which seasoned the air with their distinctive aromas.

Katya could see the sophistication of Minoan shipwrights displayed within the nearest vessel. Its deep keel, high prow, and low stern made for stability, even in rough open water. A large oak mast, perhaps fifty feet high, rose from the center with a single crossbeam to accommodate a square sail. Next to the mast, a boxy deck cabin provided shelter for the ship's captain or dignitaries.

At the back, an oar-like rudder extended into the calm waters of the harbor. What stood out to her was the ship's finish. This was less a working craft than a work of art. The entire ship's body was covered with white linen, painted with a bright purple floral pattern.

The warehouses and docks looked to be abandoned, which was perfect for them. "We're burning daylight," Xavier said urgently. "Follow me."

He scampered down the hillside, following a serpentine approach. She followed him to the edge of an open plaza, then across to a row of buildings. Xavier peered around a corner, then motioned her to follow. They ducked through a doorway. The interior was cool and dim, but Katya could make out shelves with various tools and jars. A row of pithoi vats lined one wall. Some of the vases were made of stone, others of what looked like quartz. The majority were fashioned from clay. Engravings and painted patterns on the sides indicated their contents: olive oil, grain, and wine. A crate at the back of the room held stone jars, clay lamps, and bronze daggers similar to one Mila had shown them. One of the shelves held stacks of folded wool cloth, and at the end of the shelf lay two neatly stacked piles of clothing. Xavier pointed.

"His and hers," he whispered. "Perfect."

Without preamble, he began to strip, pulling a kilt from one pile. Katya examined a blouse from the other. It seemed skimpy, like one of those you would see on some sultry actress at the Oscars. But the low cut on this blouse was more pronounced even than what the Hollywood set got away with. She turned with her back to Xavier, slipped it on, and grabbed a sort of shawl to drape over her shoulders.

In moments, they both looked the part of the average Minoan. Xavier tilted his chin toward the door and they made their exit. As they stepped into the alley, the ground shook. Earthquakes were common in volcanic regions, but the sense of unstable ground disquieted Katya.

They weaved their way between the buildings and warehouses, Xavier snapping pictures along the way. They made their way upward, out of the enclave. Xavier turned long enough to get an overview from above, and then he was off again, a man with a mission.

The hillside was mercifully devoid of foot traffic. They paralleled a path, sticking to the bushes, much as Todd had done during his Toulouse trip. Passing their pods, Katya got her first good look at Thera's central volcano. The mountain formed a raised bridge spanning completely across the lagoon. A smoldering summit caldera capped the rise at its center. Terraced olive gardens continued upslope, a stairway of black stone and green tree. Farther up the mountain, pine forests collared the great cone of the volcano. The peak reminded her of a snowless Mount Fuji with rounded ridges on the east and west sides. A flotilla of vessels bobbed in the calm Aegean waters before the land bridge, making their way to and from the beaches below them. She was standing near the top of a rugged cliff encircling the central bridge. Its three-thousand-foot-high walls stretched into the distance behind, forming a nearly complete, scalloped loop. *Strongyle.* Round. The gaps in Strongyle lay roughly at the north and south, allowing access to the Aegean and points beyond.

She could hear a symphony of birds and insects, but there was something else: music. Flute-like tones mingled with brassy percussion. Xavier didn't seem to notice. At least, he didn't slow down. They turned onto a formal roadway.

"It's right up here," he said.

Roofs rose above a tree-crowned hill beyond the pathway. Under the roofline, Katya noticed the classic Minoan design of repeating interwoven spirals. Red and black stripes ran the full length of many walls, pin striping long before race cars. As they grew nearer, she could make out voices and the strange cadence of a foreign tongue. It was a tongue that had not been spoken in three millennia. Xavier turned to her.

"This is where we need to be careful. If we can be unobserved, or blend in, it will be best. All we need is another couple minutes." He handed her the bioimager. "Remember, keep a low profile."

Xavier didn't need to tell her twice. Katya was already feeling exposed, and it wasn't just the open chemise.

A voice boomed from behind her. "Shaveeur!"

Xavier spun around. A burly Minoan rushed up from behind. "Shaveeur!" The man slapped Xavier on the back.

"So much for a low profile," Katya said.

"They have a hard time pronouncing my name." He turned to the man and began a halting sentence. The man nodded, then nodded again. Then he made a series of unintelligible sentences. Katya hoped they were not unintelligible to Xavier.

Xavier smiled as the man spoke. It reminded Katya of the way she smiled at the tax preparer's office when they explained something to her that she couldn't fathom. Smile and nod.

The man said a word and repeated it. Xavier explained, "He is saying come, come."

The Minoan, apparently a quick study, said, "Kham. Kham."

Before they moved on, Xavier put a hand on the man's substantial arm. "Duripi." He turned and put the other hand on Katya's shoulder. "Katya."

Katya smiled and bowed her head slightly as she tried out his name. "Hello, Duripi."

In turn, the man replied, "Kahchio."

"Close enough," Xavier said. Xavier and Katya followed Duripi into the courtyard. The people there were in the process of dismantling the festival.

Xavier called back to Katya, "You know what this means? We must have arrived just days after Todd's and my expedition. It's the same thing that happened in France, remember?" Xavier said. "We tried to show up early, but instead we arrived just a little bit late."

It made sense to Katya. It seemed their pods had bored a sort of tunnel in the quantum space-time continuum, a tunnel that would bring them back, again and again, with some kind of parallel passage of elapsed time.

Katya sniffed, taking in the aroma of grilling seafood, the acrid pinch of smoke. "Yummy," she said.

Their Minoan guide said, "Stifado. Octpadi stifado."

"Octpadi stifado," Katya repeated. Duripi seemed pleased.

As they crossed the plaza, Katya began to shutter images of porticos, columns, patterns in the stonework on the ground. Many of the edifices were two or three stories, with windows of assorted sizes. The windows seemed to obscure rather than define where the floors were. Grand staircases led to upper floors, with rows of bright red columns rising up the center. Two masons were in the process of mending a crack in a wall. The damage looked new.

Elegant statues stood at the center of a small square. A fountain bubbled from a central opening, lightly seasoning the air with the bouquet of rotten eggs. In the sunlight, Katya could see steam drifting from the gurgling water stream.

"A bit warmer water than the Trevi," she said.

They ascended the same stairway that had led Todd and Xavier to their feast, but Duripi led them to another building. At its base along a paved platform, a pipe protruded from the plaster wall. Steam poured from it. That was puzzling.

At the corners of the structure, horns of consecration rose, framing the top of the entire building. Double-headed axes adorned the lintel. Duripi called through the door. His train of words ended in "Shaveeur."

"Shaveeur?" came a woman's voice.

Xavier flushed. Was he embarrassed about something? Katya followed him through the darkened doorway. The door opened onto a set of steps that brought them to a sunken floor. The chamber's walls, illuminated by burning oil lamps, danced with colorful murals. Katya found the imagery astounding: the child with a monkey on her shoulder, the cone of Thera rising from the Aegean, etched by serpentine pathways. Her feelings of déjà vu morphed into specific memories. This was not similar to the mural in Mila's mansion. *It was the same mural.*

"Do you see what I see?" Katya whispered. But Xavier was distracted by the woman seated on an official-looking dais.

He spoke to her, choosing words carefully. He had been studying, and apparently it was paying off.

The woman spoke, and Xavier translated for Katya. "She says she is glad to see me again, after my strange departure. She wants to know where Todd is. Todd and I were pretty sure she saw us leave."

Xavier tried a few words, and then changed to a different sentence. He told Katya, "I'm explaining that in our travels we can only come in twos. I don't know if I communicated the point, though."

The woman spoke again. Xavier said a few more words, clearly introducing Katya. He turned to her and said, "And Katya, this is Kitane, a woman of some influence."

Katya glanced around at the temple-like surroundings. "Apparently." The Minoan woman sat on a high-backed alabaster chair atop a dais. Her cathedra was affixed to the back wall, facing toward the center of the chamber. Along the wall on either side of the chair, winged griffins faced her in bas-relief. Low benches ran along the walls, and a sunken square in the center of the room held an altar with the charred remains of what looked like vegetation. Fresh flowers had been laid across its base. Directly above was an open skylight. At the front of the altar, resting on the floor, was a porphyry basin that held water, probably some kind of sacred water for ceremonies. Katya realized that Kitane must be a sort of priestess here.

A low rumble emanated from the back of the room. To the left of Kitane's throne, a curtain opened. A man stepped out. He wore the clothing of a craftsman, with leather apron and heavy boots. He saluted by placing his right hand to his forehead. He leaned over and spoke, too quietly for them to hear. Kitane nodded and waved him to the exit.

Kitane clapped twice and said another short sentence. Xavier said, "She wants us to have refreshment." Kitane clapped again, but Duripi was nowhere to be seen. She stood with a frown and left the room, her colorful skirt trailing around her like the wake around a yacht. As she left, Katya noticed a veil of cloth braided to her hair at the nape of her neck. It was a sacral knot, confirming the woman's position as priestess.

Katya stepped quickly to the back wall and slid the curtain open. "Research, you know?" she told Xavier. She expected to see a stairway and hoped she would have time to find where it led. But behind the access was a small room with no other obvious entry. The back wall of the little chamber curved toward her slightly. Across its surface, an artist had painted a dolphin leaping from a wave. She heard Kitane's voice at the temple entry. Leaning in, she took a snapshot, and stepped back, letting the curtain close again. "Where'd that guy come from?"

She quickly shuttered images of the altar, the dais, and the ceiling.

Kitane returned with Duripi in tow, looking sheepish. The man held a tray aloft, and both Xavier and Katya took several small treats. Duripi offered them beer or seasoned milk.

Katya sampled the beer. She looked at Xavier. "Rye, I think. Or maybe barley." She took another sip and said, "How do you say thank you?"

Xavier told her, and she repeated it. This pleased Kitane. Duripi grinned, amused. She must have not pronounced it well, but it got the point across. She was beginning to pick up on the patterns of the language, catching a few words and phrases.

Xavier formed several sentences, and then turned to Katya. "I'm explaining that we can't stay, and that we appreciate the refreshments. They are...puzzled."

Kitane asked Xavier, *How do you travel? You have no ship.*

Xavier replied carefully. *We come from very far away in small...carts?*

Carts, yes.

We cannot bring things with us, or we would have brought you a gift.

It seems you cannot even bring clothing. She smiled at the Minoan garb adorning them.

We must go. Thank you, Kitane.

A whooshing sound issued from beneath the floor, and the tapestry at the back of the room opened again. Katya caught a whiff of sulfur. Two men stepped out. Katya looked at Xavier. He was clearly as baffled as she was. Kitane spoke rapidly to the two. They put their right hands to their heads, bowed, and left abruptly.

Kitane stood, watching the two leave the room. She was clearly annoyed at something, but she forced a smile. Amazing, thought Xavier, how universal some things are. She asked Xavier a question. He was uncertain of the meaning, but he feared it meant something to the effect of *May I come with you?*

Xavier tried to say, *There is not enough room.*

No, no, I do not wish to leave. I only want to say goodbye.

The less the Minoans knew of ChronoCorp technology and of people from the future, the better. But if they were to return to Thera, they must remain on good terms. Taking a calculated risk, Xavier said, *We would be honored.*

She smiled and nodded toward Katya. The three headed for the door. Katya swiveled and snapped a quick shot of Kitane, with the mural behind her. The shot would be dark, but hopefully Emil could bring out the details of the painting with his Photoshop magic.

Xavier and Katya descended the steps, followed closely by Kitane. Kitane called out for Duripi, who fell in behind her. They crossed the town square, and as they did, people along the way bowed subtly to Kitane. The party made their way down the hill quickly, picking their way through cacti, brambles, and scrubby leaves. The ghost pods were gone by now, only their indented outline left in the grass. Katya pulled out her talisman. "They're going to love this," she grinned. She triggered her fob and her pod materialized. Xavier did the same. Duripi stood with his mouth open. Kitane was more controlled, but still clearly mystified.

Xavier put his hand on Duripi's shoulder and thanked him. Then he thanked Kitane. He did not know whether to bow or to hug her. What was the cultural norm? He remembered the salute given by the men in the temple. He tried it out, and then smiled and bobbed his head. She did the same back, and said a farewell he did not understand. Xavier and Katya stepped into their pods and triggered their talismans. Kitane and Duripi stood at a discreet distance, mesmerized.

Katya had only two things on her mind. First, she had to come back here. Second, what was that strange chamber at the back of the temple? She had to find out, but that would be for her next trip. She felt herself smiling as she braced for oblivion.

41

Test-Tube Baby

The technician was sweating. "Sir, we've made quite a bit of headway, but another test of our individual pod might be preferable to—"

"Nonsense," the Ambassador said. "Use a monkey. Or better yet, send a cat. Then, if something goes wrong, we've ridded the world of another feline. Get it done by tonight. I plan on leaving for a dinner engagement at eight."

"Eight o'clock. Yes, sir."

The technician rushed back into the Transference Theater and mobilized as many of the engineers as he could find. He assigned one to phone duty, pulling in just about everyone who could make a sentence.

This time, it had to work. The equations had gone through countless recalculations. The hardware had been beefed up, retrofitted, and modified. But the smaller two-person capsule had been checked out more thoroughly than the troop carrier. The Ambassador had a fixation on the larger carrier, and that was that. Time to go for broke.

The Ambassador came in at seven. He wasn't alone.

* * *

Dresden hated the Ambassador's impromptu meetings. They often resulted in little accomplishment and a lot of discomfort. But when they called, he came. *Like an obedient cocker spaniel*, he thought grimly.

The site of the meeting was even more disquieting, considering the outcome of the last test. The Imperium's Transference Theater was the last place Dresden wanted to be.

M. Carroll, *Plato's Labyrinth*, Science and Fiction,
https://doi.org/10.1007/978-3-030-91709-8_41

"Don't look so nervous," the Ambassador said. "We're sending a rodent as a test subject. No humans this time around."

Alfie, a popular white rat in the Imperium research laboratories, had become the mascot of the technicians. When the Ambassador announced that the little furry friend would be next up in the time travel department, a communal groan rumbled through the facility. Now, Alfie sat quietly, trustingly, in his cage on the driver's seat of the tube-like Chronometric Troop Transport capsule. It was a test tube of a different kind, Dresden reflected.

The chief engineer called up to the Ambassador. "Center-point set for Italy, between 10 and 20 A.D."

"Set the controls for the heart of the sun, right?" the Ambassador quoted to Dresden. The Ambassador's normally dark countenance transformed whenever he approached the Transference Theater. He cheered up. It wasn't a good look on him. For his part, Dresden wanted to be just about anywhere else.

"Standing by," the chief engineer announced. "Interlock."

The big white tube disappeared as a watercolor in the rain. The ghost of its image lingered for a second, then vanished completely. Dresden smelled ozone and wondered if that was good or bad.

"Arrival," the tech called out. "Stage two."

"And now," the Ambassador said quietly, "for the splendid return."

The lights dimmed, just as they had during the disastrous return of the two techs in the Berlin test. The whale-like troop carrier materialized. It appeared to be completely intact.

The Ambassador leaned toward a microphone on the wall and punched a tab. "Hold on. Don't anybody move." He turned to Dresden. "Dresden, I want you to be the first to see."

Dresden forced his grimace into a smile. "Certainly," he said with mock enthusiasm. He stepped down a short flight of stairs, entered through two sets of doors, and strode across the floor to the large pod. He opened the door. Poor little Alfie was still in his cage—most of him—but in chunky liquid form.

"Well?" the Ambassador's voice boomed through the theater.

Dresden shrugged. The Ambassador motioned for him to come back up.

Dresden braced himself for a storm. The Ambassador had his back to him, still gazing into the Transference Theater. "This is becoming tedious," he said softly, then spun around and confronted Dresden. "How seriously are you taking your duties? You seem to lack the appropriate level of urgency." Though he controlled his voice, the veins stood out on his neck. "Your P.I. has one more chance. One more. I want him tapping into ChronoCorp's files by next week. Look at the calendar." He tapped Dresden's wrist and roared. "Look at your watch! One week. I'm holding you responsible." He looked through the glass at the Theater again, composing himself. Without turning, he said, "We have two active avenues into ChronoCorp, and I am getting far better intel from our inside source. The press conference is your deadline. By this time next week, I must have all we need to turn this into a success story." He turned back, locking eyes with the big Texan. "Understood?"

Dresden might have been uncertain about the Ambassador's impending success, but he was sure of one thing: PETA would not have liked this latest test at all.

42

Room at the Back

The disorienting darkness outside of the pod dissipated, and the light of the Transport Arena filtered through. Katya was home again. Todd stood over her, lifting the hatch. He held out his hand to help her out. She cradled the bio-imager in the crook of her elbow.

"You wouldn't believe—" She stopped herself. "Well, I guess you would believe. You've been there. The colors! The buildings and temples and…" Her voice trailed off as she looked toward Xavier's pod.

Brianne leaned over the open capsule. She asked, "Should we call for an ambulance?"

"No, no, no," Xavier moaned from his reclining position. "I'll be fine in a moment."

Katya felt unsteady, but she shuffled to Xavier's pod as he placed his feet on the floor. Technicians stood close by with a change of clothes for both travelers. Xavier stood, teetered, and collapsed. Brianne broke his fall by holding his arm, but he still fell hard.

Katya knelt beside him. His bloodshot eyes had rolled back. His breathing came in ragged gasps. His head swayed from side to side, and he began to mumble. "It's no good. We're running out of time. Running out. The wolves are at the door. I'll never see Thera again."

Brianne held up a phone and looked questioningly at Katya. Katya nodded. The tech dialed 911.

The first responders arrived in a remarkably short time. The staff had moved Xavier out to a couch in the lobby and tried to make him drink water.

"He's dizzy and disoriented," Katya told the EMT.

"Has he had anything odd to eat or drink? Anything out of the ordinary?"

© The Author(s), under exclusive license to Springer Nature Switzerland AG 2021
M. Carroll, *Plato's Labyrinth*, Science and Fiction,
https://doi.org/10.1007/978-3-030-91709-8_42

Katya wasn't sure how to answer that. Obviously, everything they had in Thera was odd, with a sell-by date three thousand years past. "Not to my knowledge."

The medic shined a light into Xavier's eyes. "Has he been on any travel recently?"

Katya looked at Todd. A sudden smile broke across his face and he turned away.

"Not sure," Katya said.

The EMT stood and stuffed his stethoscope into a pocket. "Definitely looks like dehydration, which can sometimes lead to confusion. But I think there may be other issues. I'd recommend admitting him."

"By all means," Katya said. "I'll go with. Todd, can you take care of the platypus and the other downloads?"

"Already on it." He stepped over to Xavier, who the EMTs had now placed on a gurney. "Safe travels, old man. I'll come visit as soon as the data's squared away from your slip."

Katya's ride to the hospital was not a Disney adventure. The ambulance was hot, and with the EMT hunched over Xavier, the back compartment was crowded. Tubing hung from the ceiling, and several monitors beeped at the side of his gurney. She would be relieved when they arrived.

En route, the EMT drilled Katya with his emerald green eyes. The guy looked like he belonged on a GQ cover. He probably spent all of his spare time at the gym. Katya didn't mind looking at him at all. He said, "So what does ChronoCorp do?"

"Research."

"Biology?"

"Physics."

"I see," he said, clearly unsatisfied. "What are you doing with a platypus?"

"Oh, that. That's a nickname we've given a piece of equipment."

"Ah, makes sense," he said, but his face wore a look of doubt. Katya let him wonder as they pulled into the hospital.

After the doctor's initial examination, and after Katya had filled out reams of forms, the physician on duty asked, "Are you family?"

"He has no family nearby, but we work closely together. We're friends." Only then did she realize the truth of what she was saying. She had considered him a friend for many years, even after he assumed the position of her supervisor at ChronoCorp. And she was worried about him.

The doctor looked at his notes, then at the form. "Well, Ms. Joshi, your friend is dehydrated, and we're in the process of treating that. He's on the verge of exhaustion. I suggest he has been sleep deprived and malnourished."

"That fits," she said, wondering if she was headed down the same road.

"There is something else to this equation. Mr. Stengel is suffering anxiety attacks. He was making little sense and exhibiting symptoms of paranoia. He was on the verge of a full-fledged panic attack. We had to sedate him. I'm afraid we're going to have to medicate him for a while. Before his discharge, we may need to discuss long-term physiotherapy."

"You think this is a permanent state?" she asked, alarmed.

"It is possible. Do you know of any family history? Is there a way we can obtain medical records for him?"

"No, I don't. Maybe if I go talk to him."

The doctor frowned. "He'll be getting groggy by now."

"Just for a minute or two?"

"Five minutes." He held up the fingers of one hand, and then gestured. "That way. Bay seven."

"Thank you, doctor."

"Bay" seven was an area the size of a closet surrounded on three sides by a flimsy plastic curtain. Most of the space was taken up by a single bed with Xavier beneath its sheets. She knew it was Xavier even before she entered: his feet stuck out over the bottom edge of the bed.

"They just can't find one of these long enough for you, can they?"

Xavier opened his eyes and smiled. "Kat, you came to visit. How nice."

"Do you remember much of the ambulance ride?"

"I think I remember..." His breathing became labored. She put her hand on his.

"Relax, Xavier. They're going to get you back on your feet in no time."

"Kat," he said in a hoarse voice. He motioned for her to come closer and began to whisper. "This has time travel written all over it. It's been getting worse."

"They can treat the symptoms and your body will heal itself. You'll recover. It's that self-healing-time-thing you were talking about."

"Oh, I don't know. I was mostly blowing smoke to appease Mila van Dijk. But what about you? Any side effects from your first trip to..." He glanced from side to side. "You-know-where?"

"I think I'm good. I'm going to study more of the language before next time. It's harder than I thought it would be, and more different than the stuff we've got in the language files."

"My notes are on my computer. Here, give me your phone and I'll put in my password."

She handed him her cell, but he couldn't focus. "I'll just dictate, okay?"

She typed in his data.

Xavier closed his eyes and said, "Before the doc comes back, what did you think of the place?"

"It was magnificent. Lyrical. Fascinating. And a little baffling."

"You mean the language?"

She bit her lower lip. "Something else. What was up with the room at the back of the temple?"

"Yesss," Xavier slurred. "The room at the back." He began to drift off as a nurse entered the room.

"Miss, it's time for you to go. You can call in later this evening and we'll let you know his room number."

"Thank you," Katya said.

"Will you need a ride home?"

"A coworker is picking me up."

Todd was already in the waiting room when she came out. "The boss is sleeping. Guess you can visit tomorrow."

Todd seemed genuinely disappointed. "I parked this way."

The two made their way to the lot. On the road, Todd remained silent for a few miles. Finally, he said, "I've been thinking about Xavier's symptoms. If we're all on the same trajectory, we might be able to head the condition off at the pass."

"How?"

"I would suggest that each time we enter a Q-slip, we take a cocktail of gentle anti-anxiety meds mixed with caffeine or some other mild stimulant to keep sharp."

"Uppers and downers together. Always an advisable combination. I still twitch from my college days."

"Not like that."

"Unless you've taken an online course in pharmaceuticals in your spare time, I'm not sure that's such a great idea. Besides, Emil warned against drugs before travel, remember?"

He let out a long breath. "Well, we've got to do something. We can't all end up as blithering idiots."

"The drugs might do exactly that to us. And Xavier's hardly that."

"But we've been watching him go down the rabbit hole. I'd hate to end up with permanent insanity." He shot her a pained smile and turned into the ChronoCorp parking lot.

*   *   *

"A triple laser?" the Ambassador marveled.

"That's what they said. They couldn't get enough power with a single ring laser system."

"Intriguing," he gushed, eyeing his monitor. The room was dark, save for the screen's eerie blue glow on his face. A technician sat beside him, taking in the conversation. "And you're sure it was a spiral?"

"She described it as a helix."

"Three lasers spiraling into time like a wine corkscrew. Increased frame dragging." He looked to the tech, who nodded. He turned back to the screen. "Makes sense."

"You're welcome. You know, Lou, you treat me like an employee."

"Technically, that's exactly what you are. You're getting paid for this."

"I'm tired of this. Tired of your attitude."

The Ambassador adopted a measured tone. "Thank you for the valuable information, and for the constructive personality critique."

He brushed a key on his laptop and the screen blanked. "I get so tired of that woman," he told the tech. "Irreconcilable differences, indeed."

"Then it's lucky you divorced her."

43

The View of the Ages

Brad found himself looking forward to Skyping with Katya Joshi. In fact, as he logged on and heard the distinctive dialing, he felt the kind of flush he hadn't felt in a long time. Not since Carrie. She had called things off after a few months, but it took him a year to get over her. He had gotten over her, hadn't he?

He realized he was feeling something else: insecurity. At the very least, he was nervous at how Katya would react to his news, which wasn't going to be as dramatic as he had first supposed.

His monitor flashed, and the top of Katya's head came into view. Her face appeared as she readjusted her laptop. A nondescript white wall backed her.

"Hey stranger," she chirped. "Funny meeting you here."

"Nonsense. I come here all the time. Are you at your office?"

The room shifted as she panned the computer around. The back of a couch came into view, and then the room opened up into a circle of mismatched furniture surrounding a throw rug. Beyond, a well-appointed kitchen displayed assorted cutlery hanging from the ceiling over an island with a sink. Past the island in the dim light stood a refrigerator and perhaps a stove and oven.

"That doesn't look like *my* office," he said. "You can actually see your floor."

"I took the afternoon off. Work has been intense. I needed a break."

"You okay?" he couldn't help but ask.

"Yeah, just wrung out. But I've been looking forward to hearing what wondrous things you found."

"Well, yes, I suppose. Look, Kat, I'm afraid I may have spoken in more extreme terms that the situation warranted."

M. Carroll, *Plato's Labyrinth*, Science and Fiction, https://doi.org/10.1007/978-3-030-91709-8_43

She squinted at him sideways. "Where did all that boyish excitement of yours go to?"

He shrugged self-consciously. "Truthfully, the changes are less than I thought. When I first laid eyes on several of the sketches, I thought the guy had taken a course in modern paleontology. But upon further examination, the differences are less marked, less obvious. It feels like someone came sneaking in during the night and tweaked the drawing. Added something here, erased something there. Are you…disappointed?"

"No, not at all."

"There's something else: I can't find the shots of the images that I took with my phone. The files disappeared. Maybe I dumped them accidentally, but I really don't think so. If you ask me whether the drawings have changed, I'd have to say it appears so, but I can't document the fact."

He watched Katya sitting there, saying nothing. He gave her space to think. Suddenly, she yelled, "Wellington!"

Brad thought he was up on all the new expletives, but he wasn't familiar with this one. Katya disappeared from the screen, but he could hear her voice, scolding someone. She appeared again with a soggy sponge in one hand and a cat in the other. "Wellington, meet Brad. Brad, this is Wellington. I think he's jealous. My coffee was two days old anyway."

"Maybe Wellington is just a connoisseur of finer java."

Katya became quiet again. Her stare was defocused as she scratched the kitty's head. Even through the computer, Brad could hear him purring. Finally, Katya said, "Well, I'll have a chat with the powers that be, and then I think it will be time to get you out here for a tour."

"Inner sanctum. Are you sure I won't be struck down by the quantum gods?"

"These days, I'm not sure of anything." She said it cheerfully. "Thanks, Brad. I'll be in touch."

* * *

The last time Katya visited a hospital for any length of time, it was because she had contracted a nasty coronavirus. She'd gotten off easier than many: she was in bed for a dozen days. Entering the ultra-modern Skyline General Hospital gave her the shivers on principle. But the staff put her at ease, and when she finally made it to Xavier's room, her stomach had settled down.

Xavier was sitting up in his bed, computer on his lap, propped up by a mountain of pillows. He continued to stare out the window as Katya entered and said, "Hey Boss. Looking good. How are you feeling?"

He continued to face the window. "Mila stopped by this morning."

"Did she, now?"

Xavier didn't elaborate. Instead, he said, "You know, Kat, all of my professional life, when I wasn't up to my eyeballs in theoretical physics, I've studied dead places, failed societies, beautiful things that came to an end."

He blinked slowly and turned to look at her. "Ironic, isn't it? To be alive in one time and yet spend all your mental energy living in another?" A wan smile flickered across his face. He closed his eyes and leaned his head back against his pillow. Katya studied him. He seemed deflated, as if the last trip had dissolved some of his spine. Perhaps, she thought, he was facing something he had never faced before: mortality. The concrete fact that all things have a shelf life, that all things—whether people or pets or civilizations—must eventually come to an end.

He opened his eyes again. "Nice of you to come visit, Kat."

"Well, you know, I don't have anything to do at work because my boss is gone. You'd be amazed what people can get away with."

Xavier chuckled. "I'm sure. By the way, I scheduled our press briefing. Three weeks from tomorrow."

A hot wave of exasperation washed over her. "I thought we were instigating a coordinated approach. Isn't that what you said?"

He waved the comment off. "Try to enjoy yourself. You've seen things no one else from our time has. Amazing things. The press and our colleagues will want to know. We have two papers accepted into journals already. Time to pull the trigger. They say they're letting me out in a day or two, so I'll be back in plenty of time to make party favors. But if something goes wrong, it's up to you. You do a great job with the press, and you're far better at the Q&A than Todd or me."

"Thanks, Boss, but you won't get off that easy. I'm counting on you being there."

In the parking lot, Katya sat inside her car, fuming. She would need to get that list of potential invitees Todd and Emil had come up with, and have someone—maybe Brianne—send out evites. They would need a banner as a backdrop for the podium at the front of the largest meeting room they had. Where had they put that podium? And she would need to personally invite Bradley Glenn. Was that too weird? No, it was the thing to do.

* * *

Bradley Glenn opened the email from Katya. The subject line simply said, *Just for you.* "Well, that was weird," he mumbled. He clicked on the link. A personal invitation to ChronoCorp's press conference appeared. He looked at the date. Then he looked at his calendar. He had to call Katya.

The phone rang. It rang again. Kat picked up.

"Katya, I'm intrigued by the ChronoCorp press conference, but as luck would have it, our 'Hadrosaurs of Hudson River' is opening just half an hour before your meeting."

"There's that synchronicity thing again."

"Great minds, Katya. Great minds."

"I'm sorry you can't make it, Brad." She sounded genuinely disappointed, so he forged on.

"But how about if I come out in the afternoon to give you moral support? I checked flight times. I can be to Denver International by one-thirty, up to Fort Collins by three?"

"I like that. If things don't go well at our meet, you can take me out for a few margaritas."

"Deal." As he hung up, he realized that he was looking forward to Colorado more than his own event.

44

Facing the Music

Rex Berringer was done with sneaking around. This time, he would confront the personnel of ChronoCorp face to face, mano a mano.

Wearing a plumber's suit.

It was his most recent run-in with Dresden that supplied his inspiration. The man spoke to Rex as more of a colleague than a malevolent overseer, and that was a big change. Something had happened. Dresden said ChronoCorp was doing big and important things that the Imperium needed to know about, things the Imperium was also attempting. Being the talented investigator that he was, Rex had gleaned that much. But then Dresden had mentioned something about bending space and time. Rex didn't know a lot of higher math, but he had seen *Back to the Future*—the entire trilogy and the blooper reel. He had also figured out—all by himself—that it was probably a dangerous thing. That was even before Dresden made the point. So ChronoCorp and the Primus Imperium were both up to no good, and Rex was up to his eyeballs in it. But he wouldn't get paid, and he wouldn't be able to extricate himself, until he finished his assignment at ChronoCorp.

Rex didn't know what kind of education most private investigators had. For his part, he studied theoretical economics at Nebraska State with his buddy, Randy. Rex's practical application of said economics was abstract at best, while Randy had actually taken part in the economics of the land: he had become a pipe fitter, and he was raking it in. He also owed Rex several favors, and paid them off summarily by forging some official-looking plumbing inspection documents for him. He had also loaned Rex a hat.

This time, Rex knew someone would be there at ChronoCorp, because they had some press conference shindig scheduled for all morning. That was

M. Carroll, *Plato's Labyrinth*, Science and Fiction,
https://doi.org/10.1007/978-3-030-91709-8_44

fine with him; while putting their best foot forward for the press, their guard would be down.

The press conference was scheduled for 10 am Mountain Standard Time. Members of the press corps would begin filing into ChronoCorp by 9. Rex waited until a good amount of distraction built, arriving at 9:30. He announced himself at the desk as plumbing inspector Kelsey Acres, setting off a flurry of activity. Overhearing this, an engineer named Todd approached the main desk, where he asked the receptionist to contact Xavier. Rex knew the name Xavier Stengel. He was the big kahuna, the main cheese, the boss man.

Xavier appeared within moments, Brianne in tow. He glanced at his watch, a worried look on his face. In fact, the boss looked a little haggard and pale. "The county just did that last month," Xavier said, as if Rex wasn't there. "What do they want now?"

"Not the county," Rex said, pretending to study his old-school clipboard. "I'm from Tri-county health."

Todd pointed at the clipboard. "When are you guys going to catch up to the twenty-first century?"

Rex ignored the comment. "Your laboratory is doing exacting work, and that requires exacting supervision." He peered over his clipboard, casting an extra look at Brianne, who fidgeted uncomfortably. "Exacting."

Xavier asked to see the man's identification. Rex handed over the clipboard. Xavier studied it for a moment. "Kelsey Acres. Well, Mr. Acres, I suppose you'll want to examine the main and secondary lines in the lab?"

"That will be fine."

Xavier turned to the tech. "Todd, could you please show this gentleman to the mechanical room adjacent the labs?"

"Sure thing."

Xavier added, "and have Brianne keep him company during your part of the conference, will you? In case he needs anything."

At the mention of her newest babysitting duty, Brianne gasped audibly but held her tongue. A woman behind the desk asked Rex to leave his cell phone with her.

"Really?" he bleated. "I mean, really?"

She nodded and held out her hand. He gave it to her, squinting through his thick glasses, part of his perfect disguise.

ChronoCorp's bigwig disappeared down the hallway. Todd turned to Rex. "Now, what all do you need?"

Rex was glad Xavier had mentioned the mechanical room. That sounded official and technical. "I'll need to see the mechanical room."

"Plumbing or electrical?"

"Plumbing. Just have a few forms for you to fill out, but I don't have them in hard copy. I'll need to hop onto your system." Rex tore a corner from one of the official-looking forms on his official-looking clipboard. He prepared to scribble something on it.

"To our system?" Todd asked.

Rex brushed the tip of his pen to his tongue. It tasted surprisingly awful. He kept his professional cool. "I just need a password so I can log on and access the company online forms. I'll download our files for you to sign."

"I don't think that's a very good idea," came a voice from behind. A tech wearing an "Emil" nametag had come up to the desk, scowling at both of them.

Brianne gestured her colleagues aside, then said in a hushed voice—but not quite out of Rex's earshot—"I know this guy."

"How?" Todd blinked.

She made a face. "From high school."

"Did you two…date?" Xavier asked.

"No, no!" Brianne soured. "Well, almost once, but no. The last I saw him was just before prom; I ended up standing him up. He was no catch then, and apparently not much of one now, either. Not exactly the brightest bulb…"

At this, they all glanced back at Rex, who stared dumbly at the ceiling, humming to himself. Sure, he would play the brain-dead, bored employee; it was a role at which he excelled.

"I'll sign him on using my password," Todd finally conceded, loud enough for all of them to hear. This suited Rex well enough: he could get files onto his small-but-mighty flash and be on his way just the same.

Todd escorted Rex to the room with all the tubes. As Rex tried to look interested in the pipeworks, Brianne entered, keeping a noticeable distance between them. Todd left Rex with Brianne and headed for the conference.

As soon as he had gone, Rex whispered, "Fancy meeting you here."

* * *

Katya scanned the meeting room. The turnout was surprising; Xavier had hoped for a handful of staff members from Internet services and maybe, if they were lucky, a blogger from some major outlet like the New York Times. But as Katya looked over the signups, she realized that they had made a splash already, despite the fact that nobody knew what they were really up to yet. Several print reporters, one from a major archeology magazine, were in attendance. A technology commentator from the BBC was also there, as were several representatives from tech companies and universities. The moderator of the popular CNN show *Science and Life* sat in the front row, and several scientists of note also joined in. Most notably, Alexander Malling had flown in from the Oriental Institute of Chicago with two colleagues. Katya remembered the name: he had been a thorn in Xavier's side for ages, though she didn't know the specifics. The room was maxed out; they had to send Brianne off for more coffee.

Katya watched as Todd and Emil conferred. Mila van Dijk walked into the room, and her entrance surprised Katya. It wasn't the woman's formidable presence or her self-possessed demeanor. What brought her up short was a subtle nod Mila gave to an unlikely person: the plumbing inspector. If the guy had shown up on her doorstep for an inspection recently, it would have been a remarkable coincidence. Too remarkable. They had met before.

Todd disappeared with the plumbing expert but soon returned. Once he was back, he gathered Xavier and Katya for introductions. The lights dimmed, and his presentation began. An image of the Great Pyramid of Cheops shimmered across the screen.

"The Egyptians were arguably the greatest power of the ancient world, circa 1500 B.C. They ruled the roost, and they had little time for outsiders. The Egyptians referred to foreigners as 'Barbarians who are an abomination to God.' On a good day, they called them Ha Unebu—'People from beyond the seas.' And while they cast aspersions at all the tribes and people groups in the region, they made one exception: the Minoans. The ancient Egyptians referred to the Minoans by a separate name: the Keftiu. Such was the great respect the Egyptian empire had for the sea peoples who came from the Aegean. It is no wonder: The Minoans ruled the open seas, with ships far superior to any of the time. While the pharaohs trundled up and down the Nile in their small barges, the Minoans ruled over the Mediterranean not with military might, but with trade and economics. Would you like to see a photo of one of their ships?"

Todd paused for effect, and then brought up the best image they had of the quay he and Xavier had visited. "This is not an artist's digital reconstruction. It is not an archeological site or a museum display. The ships and docks and people are real. This photo was taken on site in ancient Thera in about 1600 B.C. Give or take."

Nervous rustling filled the room. A few hands went up.

"Before we get to your specific questions, let us outline our approaches and goals, and our results. For the technical side, I'd like to introduce Katya Joshi, who…"

Todd's voice faded away as Katya braced herself for her part of the presentation. She stepped to the podium and flooded the screen with charts and graphs. The journalists fidgeted in the dark.

Although Katya's material might have seemed dry to some, her forceful presence kept their attention. She exuded confidence in her subject and held herself in a way that communicated that she was sure of her data and herself. Once Katya had overwhelmed them with the technical side of ChronoCorp's time travel technology, it was Todd's turn again. He had the advantage of visuals.

"The period of the 'old palaces' spanned from 1900 to 1700 B.C. That's when the large central palaces began to take shape: Phaistos, Malia, Knossos. But at the end of that phase, moving into the Neopalatial period, these structures were practically obliterated, probably by tectonic forces like earthquakes. Then came the new palaces, built on top of the ruins of the old. For the first time, we have the chance to visit and document the structures of the old palaces, structures that remain a mystery today. But for now, we concentrate on the Minoan civilization of the 17th century B.C., that of Crete and, in our specific case, Thera. What we saw was wondrous."

Todd advanced the image with a melodramatic gesture. The research community had already seen a release, hours before, of some of the images. But ChronoCorp had saved a few of the best for now. It was a feast for the eyes: Minoans in the city square, in the full regalia of Minoan fashion; ships coming into harbor from Crete or farther; architectural details and murals, sculptures of goddesses holding snakes (or were the snakes holding them?).

"Are you suggesting," one of the physicists piped up, "that we are seeing photographic imagery from thousands of years in the past?"

Todd knew the question would come, despite all the work they had done ahead of time to quell such doubts. And he had hoped Katya's presentation would preclude most of the doubts. "Absolutely. And we will get to more of the technicalities, the engineering behind these images, in a few minutes. There will also be a separate press conference on the imager we used. That will be scheduled soon."

Voices filled the room. With Todd's next images, the journalists quieted.

"And here we see the classic outline of a Minoan complex: long, straight outer walls oriented to the north, with a central court surrounded by a warren of small chambers." With each image, Todd pointed out specifics in architecture, city planning, and even shipbuilding.

So far, so good.

45

Snooping

When Brianne and the inspector arrived at the mechanical room, the man seemed to be having trouble seeing. He whistled some eighties disco number, peering over his thick glasses and clasping his hands behind his back, leaning toward the various pipes as if they were dangerous. Apparently, he didn't see the control box. She cleared her throat.

"Did you want…" She nodded toward the box.

"Ah, yes, there it is. Certainly." He stepped over to the metal access and yanked on the handle. The metal screeched, but remained sealed.

Brianne pointed to a latch on the top. "I think you need to undo that part first?" She offered.

He grumbled something, fumbled with the latch, and opened the little door. He peered inside at dials and monitors, and scribbled some notes on his clipboard.

"So, Rex, it's been a long time."

"Yeah, yeah it has. I imagine we're both different people these days."

Brianne's phone chimed. She studied the screen. The Q&A portion of the press conference was in full swing, and it looked like they were having a sound issue. She was trained in audio, but what about the plumbing inspector?

"Shit," she whispered.

"Do you need to go? I can finish up here. I'm almost done. You know you can trust me." He said the last with dramatic emphasis.

Brianne looked right and left, wishing she had someone to ask. She needed to see how the conference unfolded. "You have to go directly to the front desk when you're done here. Promise me."

M. Carroll, *Plato's Labyrinth*, Science and Fiction,
https://doi.org/10.1007/978-3-030-91709-8_45

"Look, I don't want to do anything that would get you in trouble. I'll just be in and out."

Brianne turned toward the door. "Don't forget to sign out when you turn in your badge."

"Will do. Thanks for all your help."

Brianne raced down the hall. She was needed, and now was an opportunity to demonstrate what a valuable employee she was, to get in their good graces, to gain their trust.

*　*　*

Rex had no intention of getting Brianne into trouble. His intention was to snoop around and find answers. It was time to put into action the *investigator* part of *private investigator*.

He accessed the mainframe, but couldn't find anything specifically labeled as "time travel notes" or "how to make a time machine," so he had to wing it. First, he plugged his flash into the system and downloaded any files associated with the engineering of the place: blueprints, lists of equipment, even spreadsheets of purchased hardware. His phone was still at the front desk, but he had a small camera in the lining of his jacket, as any respectable P.I. on TV did. He would use it if he saw anything interesting, and even though the place dripped with boring science, there was bound to be something of interest in some dark corner, somewhere.

When he felt he had the quintessential *Idiot's Guide to ChronoCorp* on his drive, he pulled it, stashed it in his coat, and tried to create a virtual bridge from ChronoCorp to the Imperium. Something didn't quite add up: he could connect, but only sporadically. A little red shield kept appearing in the corner of the screen. The place had exquisite security. But he had what he'd come for, and now he would get as much of it out as he could. Using the workstation Todd had so generously given him access to, he transmitted a handful of important-looking files to the secure Imperium site—as many as the system would allow before it threw the little red shield at him again. He logged off and turned away from all those baffling electronics.

Rex Berringer, P.I., stuck his head through the doorway. He pulled his thick glasses off, peeked both ways, and put them back on. The place was poorly lit. ChronoCorp must have been largely abandoned for the press conference, just as he had hoped. After one more glance, he stepped into the corridor in search of a company directory.

*   *   *

The early questions were softball queries. Todd overheard a woman sitting in the front row as she leaned over and spoke to the reporter next to her. "What's with all the snakes?"

Todd noticed her ID badge: she was from the blog Colorado Physics. He said amicably, "Snakes were not seen as threatening. In the Minoan belief system, they were seen as benevolent visitors from the netherworld. Have I got that right, professor Malling?" His question brought the spotlight on the important archeologist and gave the man an opportunity to show off a little. It seemed a good strategy for cozying up to him. Surprisingly, the scholar even remembered Todd's name.

"Yes, that's correct, Dr. Tanaka. Serpents were an important part of the Minoan construct. They're often seen in the hands of a goddess figure prevalent throughout the culture."

But as Todd, Xavier, and Katya began to relax, the torrent began. Among the physicists, skepticism seasoned their questions. One journalist accused ChronoCorp of spinning fake news. One of Malling's colleagues stood and flared, "Stick to physics and leave the archeology to the experts." The veins were sticking out in his neck, his face red. Xavier bristled.

When Katya called the proceedings to a close, several researchers left the building in the midst of heated arguments with each other. ChronoCorp had certainly generated controversy. Xavier sauntered over to Todd, keeping his eye on an approaching figure in a sharp business suit.

He whispered "Stay sharp for this one, Todd." Xavier shoved his hand toward the man. "Dr. Flenner, welcome."

"Dr. Stengel. It's been awhile," he said neutrally.

"It has indeed. This is Todd Tanaka, as you know, and Katya Joshi."

"Pleased to meet all of you," Flenner said, not taking his eyes off of Todd.

"Dr. Tanaka, are you by any chance familiar with my work?"

"Certainly, Dr. Flenner," he lied.

"It turns out that one of my areas of study has been the plumbing and drainage of the great Minoan palaces. My particular emphasis has been on the Palace of Knossos. It's a huge complex, as you well know. Handling water throughout such a large collection of structures is a daunting challenge, and it is a challenge that the Minoans met. They built a network of cisterns, aqueducts, wells and so forth." He jerked his chin toward the darkened screen. "You had several detail shots of roofs. Of cornices and downspouts and catchments. They reveal details that are quite technical and not well known in the literature."

He cast his gaze around the room and spoke confidentially. "I don't understand the physics behind what you've done, but I know my ancient architecture. Congratulations." The man left quickly.

Suddenly, as if she had materialized from thin air, Mila van Dijk appeared. "This was a bit more colorful than I expected. I once attended a lecture in Amman where a scientist actually took a swipe at one of his colleagues. It was half-hearted, of course." She smirked, examining the room as the last attendees departed. "Thanks for the entertainment. For what it's worth, I thought you all handled yourselves admirably."

Todd began to relax for the first time the entire day. But the aftermath of the press conference wasn't the only storm brewing in Fort Collins.

46

Strange Rooms

Funny, Katya thought, how differently people react to the same stress. Xavier was practically giddy in light of Flenner's comments. Todd fumed at the overall reception their hard work had received. Katya was exhausted, but not too tired to lick her wounds. She had anticipated some skepticism, but not the outright insults from colleagues in the sciences. It might have helped to give the press corps an abbreviated tour of the facility, but Xavier had been adamant about security concerns.

With Bradley due at 3, Katya kept finding excuses to wander into ChronoCorp's lobby. The receptionist was too polite to comment. Finally, Katya sat down in a chair and pretended to page through a magazine.

It was a busy afternoon. The chime of the elevator kept a steady pace. She watched as the people stepped in, disappeared, and others came out. It reminded her of something. Something important. She was so tired. The adrenaline rush of the morning was dying off.

She rubbed her eyes, stood, stretched her back and stepped over to the coffee machine. She poured herself a cup and took in the energizing brew. With a start, she remembered what had struck her as wrong about the room in Thera. When the workman had first come through the door, Katya noticed the painting of a dolphin on the back wall. But unless she was mistaken, when the others came through, it had disappeared. She didn't register what it had been replaced with, but it seemed something innocuous like a floral pattern. She would have to get Xavier alone and compare notes. The important point was that the entire room behind that curtain had changed. People had come in and out, just like the elevator she was staring at right now.

M. Carroll, *Plato's Labyrinth*, Science and Fiction,
https://doi.org/10.1007/978-3-030-91709-8_46

She realized that someone was standing in the elevator doorway, saying something.

"Earth to Kat. Am I late?"

It was Brad. His hair was tousled and his clothes bedraggled, but somehow she didn't care.

"It's nasty out there," he said, shaking some rainwater from his sleeve. He came in and shook her hand. It was awkward. She wasn't sure if a hug was more in order. Were they friends? Colleagues? A little of both? The line was as fuzzy as her brain felt.

Kat looked through the window. Bradley had arrived beneath steely purple skies. Toward the east, a patch of blue peeked through, but lightning laced the ragged edges of the clouds.

"Looks like you got a little precip on you," Kat said. "Typical Colorado. Plains bake in the morning, and in the afternoon the clouds boil up. My Dad calls it God's air conditioning."

Bradley leaned toward the window and grimaced upwards. "That's pretty respectable rain all right."

"Sometimes we get hail," Xavier said, approaching to shake his hand.

"Sometimes we get tornadoes," Todd called as he came up the hall. "Hey, Brad."

"Hi Todd. Wouldn't want to be up here in one."

"Yep, tallest building on this side of Fort Collins. But there's a basement."

Bradley rubbed his hands together. "I've really been looking forward to seeing the Inner Sanctum. I suppose you throw a switch and sparks fly and liquids bubble?"

"Better than Frankenstein's laboratory," Kat teased. She looked at him over her shoulder, immediately realizing it could be mistaken for a flirtatious gesture. She fought to recover a neutral posture. She hoped she wasn't blushing again, but her ears felt a wave of heat. *Stoplight Katya*.

"I was thinking more along the lines of the old TV show with the two guys who lurched down this long tube," Brad said. "Timetube? Time Path? Lots of blinking Christmas lights and levers to throw and buttons to push. All it took was the right button, and you were off on your trip."

"Time Tunnel?" Xavier offered.

"Send the men to the deck of the Titanic," Kat quoted from the show.

"Right. I didn't think it would be that easy," Brad admitted. His foot slipped and he nearly fell to the floor. Kat caught his elbow and pulled him up. Their eyes met briefly.

"You okay?" she asked.

"Yeah, it's these damned shoes. They're new. Got 'em wet outside and they're slicker than snot."

Xavier went on as if nothing had happened, smiling at Brad. "Of course, it turns out that that's exactly what we do. Not with a switch but with a touch screen."

"Remarkable."

"The application is remarkable. But quantum physics can be defined by a host of equations, just like Newton's math showed us how gravity works; how things move."

"Yup," Todd put in. "Newton's arithmetic defined how the visible universe operates. The math in the quantum world—the equations of the subatomic realm and the interplay of space and time—is weirder. It's beyond bizarre. To some, it suggests multiple realities. And any observation you make appears to select one of those possible realities. The other realities, if they were ever there, either disappear or remain hidden somewhere. So either we can't move freely through time, we can't change it, or we cause a divergence of possible universes."

"Well, which of those is it?"

Kat said, "Depends on whose math you use."

"So nobody knows."

Xavier met Brad's comment with a pensive smirk. "When it comes to quantum physics, you might as well get used to disappointment. And weirdness. It's all pretty counterintuitive. Coffee?" The three made their way to the conference room.

Katya picked up where the boss left off. "Niels Bohr said reality can only be explained in terms of seemingly contradictory concepts."

"Complementarity," Todd said around a cinnamon roll. He held one up to Brad, who declined.

Katya nodded. "Something like. For example, light is a wave, yes, and it's also a particle. Not either. Both."

Brad said, "So, something like my former tax accountant. She was a bookkeeper, and she was also an idiot."

"Which must be why she's your *former* tax accountant," Kat said.

He held up his foam cup. "Exactly. Hey, I caught part of the Q&A."

"That was the most, er, stimulating part, for sure. How are the Hadrosaurs of the Hudson?"

"They were a big hit. 'Dead beasts come to life,' one reporter called it." He shrugged. "And that was fine, I suppose. We got good coverage, and first-day attendance is high."

"Congrats."

He peered toward the doorway. "Hardly compares with what you're doing here, though."

"Can I get you something to munch on? Donut? Cheese puffs?"

"Tour?" he suggested.

She laughed. "You've been patient long enough."

* * *

ChronoCorp offered all sorts of mysterious treasures. Rex found his way to the electrical room, but its workings were beyond him so he moved on. Farther into the bowels of ChronoCorp he discovered the "Lab," a room with lots of computers, countertops with bizarre experiment-looking thingies, and a row of gelatinous, meaty blobs the size of a human heart. They seemed to pulsate and even breathe in a way, and each looked like it had a toupee slid off to one side. He had no desire to get too close to the slimy objects, some of which were attached to a tangle of electrodes. He could have sworn one of them blinked at him. He took lots of pictures, hoping they were more than just somebody's leftovers from a bad lunch.

Then, he made the discovery that was the high point of his day: the employee lounge. He grabbed several flavors of donut holes, a napkin for the powdered sugar one, and half a cup of tepid coffee. Cool coffee was better than no coffee, he always said.

He considered: should he go on, or go home to safety with the booty he'd already accrued? To the right lay the exit; left was uncharted territory. He paused in the doorway, turned left, and passed several closed doors, either locked or unimportant. He could hear a sort of buzzing coming from farther down. There, light poured from an open double door. Above the door, a sign declared TRANSPORT ARENA. Now *that* sounded important. In fact, the doors had a slot for a key card, so the room was supposed to be secured, which made it important. But somebody had not quite shut the door. Rex was grateful. He stepped through.

The place wasn't nearly as dramatic as its sign suggested. To Rex, it looked like a tanning salon, except for several computer stations scattered around the room. Duct tape secured snakes of cabling to the floor. The tanning booths themselves were of a style he had not seen, although he didn't spend much time in places that made you sweat voluntarily. On second thought, he was fairly sure these were not tanning booths at all, but what were they? These objects reminded him a little of coffins with miniature jukeboxes on the ends. He figured they might be important items of hardware. He pulled out his little camera and prepared to take a few shots, but a thought was nagging away at the back of his mind.

Finally, it came into the clear: between ChronoCorp and the Imperium, had he chosen the right side?

47

Thunder and Lightning

Xavier reminded Brad of the confidentiality contract Katya had given him. Katya thought it unnecessary, but Xavier was the boss, and she wasn't about to get into a security discussion with him now. Once that formality was out of the way, they were off. Xavier let Katya take the lead, which relieved her. Apparently, he was feeling much better after his hospital stay, but she wanted to take Bradley Glenn around the building complex with all its marvels.

Katya began with the heart of the place, the foundation of their experimentation. She tapped on the metal plating at the back of a nondescript room.

"Behind this wall is a ring laser, the key to our travel. In fact, multiple lasers. At the back of this building, three powerful lasers spiral up a tower. They set up something called frame dragging."

Brad looked lost, so Xavier tried to clarify. "A physicist named Ron Mallett first came up with the idea."

"Of frame dragging?" Brad asked. Katya could tell he was trying to keep up.

"Not of frame dragging specifically, but of using lasers to trigger it. He said that time is moored to space. We know that gravity is essentially the warping of space; we've known it ever since Eddington watched starlight bend around the Sun during a total eclipse. That was in 1920 or so. But gravity also warps time. Mathematics tells us that time slows as a traveler approaches a black hole and accelerates in speed, for example."

"Too bad it doesn't work that way for highway speeding," Brad quipped. "I just got a ticket from a nice State Trooper on I-25. If only time had slowed down enough for me to get off the road."

© The Author(s), under exclusive license to Springer Nature Switzerland AG 2021
M. Carroll, *Plato's Labyrinth*, Science and Fiction,
https://doi.org/10.1007/978-3-030-91709-8_47

Xavier smiled. "You just weren't going fast enough. So now we have this simple set of relationships, suggested by Mallett: if light can affect gravity, and gravity affects time, that means light also can affect time."

Katya jumped in at a level she thought a non-physicist could understand. "Imagine our lives as a line stretching from the past to the future. Time and space are connected to each other. If we can twist space strongly enough, we are able to twist time as well. That time twist becomes a loop that we can jump into."

"What Mallett was talking about," Xavier said, "and what we've done here, is to set up a powerful ring laser. We tried a classic ring at first, but couldn't get enough power."

"Yeah, power has always been the problem," Katya said. "How do you channel enough energy into a confined space? We did some quantum magic to get ours to work. Patent pending."

"Right," Xavier said. "So one strong ring laser just didn't do the trick. The solution was to create a spiral of three lasers. As light passes along this spiral, the space/time in its center is 'stirred'. The ring laser swirls empty space like a rotating spoon stirs molasses. It twists the time that we perceive as a linear affair, folding it around into that loop Katya mentioned. It's called a closed, time-like curve. This disruption of the space/time continuum is essentially our portal to other locations on the timeline. We simply have to learn to control the size of that loop."

"Lasers opening a doorway to time." Brad looked around the room and whistled. "Television meteorologists would have a field day with this technology. Talk about accurate weather. 'Forecast tomorrow, rain beginning at 4:17 pm.'"

"That might fall under the category of abuse of power," Katya said playfully.

Brad became pensive. "So, no Time Tunnels. No Morlocks."

"They make a good story," Xavier said. "The lovers who leave letters in a mailbox that bridges the century. The physicist who climbs into his DeLorean and drives off into the past. In reality though, time isn't so tidy. It drifts and shifts in currents and eddies. From our standpoint, the river only flows one direction, but once you're in the stream, these quantum subtleties can take you in unpredictable directions, in both space and time." Xavier seemed proud of himself, but Todd was frowning. "Did I get something wrong?" Xavier asked.

"Not wrong, exactly. Just imprecise. It may be more like a slinky than a loop, where it doesn't loop back on itself, but rather swings back into a slightly different place in reality. A different universe, possibly."

"I'd read about this," Brad said. "So we might have a million universes, each reflecting some little decision we make. Sounds like a mess."

Xavier rethought. "I lean more toward a notion called decoherence. Fascinating idea. It says that we have several possible realities, but only some of them are stable and permanent. Versions of reality that are robust, ones that are strongly imprinted on the environment, survive to be observable. The other ones just dissipate."

Brad frowned. "So you have a bunch of universes that just sort of dissolve away?"

"Just like that Hawkins drawing that's no longer with us."

Kat gave a start. "No longer with us?" she asked Xavier.

"Chad Markham emailed. Evie Long reports that the drawing is now white powder. I guess it didn't make a big enough impression on the universe to hang around. Good news, yes? We can do what we want and when we want, within reason, of course. No repercussions."

"I suppose so," Kat said, forcing some joy into her voice. But she began to wonder: would anything make a difference? Could she make a difference in the way she planned?

"So," Brad said tentatively, "maybe the changes I saw—or didn't see—in the Hawkins drawings were because they weren't imprinted upon the universe enough to stick?"

"That's the idea," Katya said. "Not bad for a paleontologist." Katya moved toward the door. "And before we show you to the place where we depart, we'll pass by what is a surprisingly fascinating part of the tour: the plumbing. It takes a lot of liquid to cool those lasers."

"Speaking of plumbing, may I use your restroom?" Brad asked.

Todd and Kat locked eyes and whispered in unison, "The Time Chamber."

Outside, thunder rumbled ominously.

48

Piddles and Puddles

The Time Chamber wasn't what Bradley Glenn expected. Its door had the universal sign for restroom. Katya couldn't remember who came up with the term "Time Chamber," but the facilities were labeled as such with a clean, official-looking sign. She alleged the name fit the temperament of either Todd or Xavier. "Its genesis is shrouded in the mists of time," she said theatrically.

Brad stepped into a stall. Its metal partitions were plastered with signs of all shapes and sizes. A weathered wooden placard with a picture of a frog sticking its tongue out read: "Time's fun when you're having flies." Above the door, a poster declared, TIME IS OF THE ESSENCE. There were others.

"Time is what keeps everything from happening at once."

"One thing leads to another."

"When This Baby Hits 88 Miles Per Hour, You're Gonna See Some Serious S***."

"A stitch in time saves nine."

"Time is money."

"Time is the wisest counselor of all."

"Better three hours too soon than a minute too late."

"Got time on your hands?"

"Time flies like an arrow, but fruit flies like an apple."

As he washed his hands, the lights flickered. He looked up to see a sign above the mirror urging him on. It showed a round-faced kid with his cheeks squished between the doors of an elevator. The caption read, "Pressed for time."

Brad met his trio of tour guides in the hall. "What was that with the lights? This storm?"

© The Author(s), under exclusive license to Springer Nature Switzerland AG 2021
M. Carroll, *Plato's Labyrinth*, Science and Fiction,
https://doi.org/10.1007/978-3-030-91709-8_48

Xavier looked frustrated. "Yeah, but it's not supposed to happen. We take measures here, with all the delicate equipment. There's no telling what an ill-timed power spike might do to our experiments."

"Or our coffee maker, for that matter," Todd added. "Hey, maybe now's the time to pray for an espresso machine. One nicely aimed bolt, and pow. We need a new brewer."

"Shall we see the Transport Arena?" Kat asked cheerfully. It sounded better to Brad than the Time Chamber. They made their way down the corridor toward the double doors.

*   *   *

Rex prided himself on his artistic framing of all his photographs. His last client had told him that his photography was "creative." Oddly, she had fired him soon after.

The wall before him was puzzling: a row of four necklaces hung by a few tattered bathrobes, clearly hand-me-downs. What did they need bathrobes for? A handwritten sign above proclaimed, "The hallowed robes of return."

Rex had just imaged the jukebox assembly on the end of one of the tanning booth thingies when he heard laughter and voices coming down the hall. As the jovial people drew closer, he realized they might be on their way to this very spot. The place had no closets, no big pieces of furniture, no curtains. In fact there was only one place in the room to hide: a tanning booth. He chose the closest one, rolled into it and slammed the top down. Strange how no lights came on.

*   *   *

"And as you can see," Katya narrated, "we've got two pods. This is for redundancy, but also so we can send more than one person at a time."

"But no big levers, no sparking coils," Brad complained. "You guys just didn't decorate your laboratory right. Still, it's impressive."

Xavier leaned over to a monitor, brought up some readouts, and pointed. "This takes the place of all our Frankenstein props."

"More's the pity," Kat said. "It would be nice if we at least—"

The lights flickered again. "That's not good," Todd said. "I'll go see." He stepped quickly to the door. Brad got the impression he took off at a dead run as soon as he was out of sight.

Xavier continued, tapping the touchscreen. "These readouts here tell you a location in both 3D space…"

"And also location in time," Katya added. "You have to think in four dimensions here. And locating a target along the time axis has been the most difficult."

"I can imagine. And I listened in on some of the press conference. Thanks for the link, Kat. So…ancient Greece, right?"

"It's been amazing," Katya grinned. "And now we can actually talk about it."

"I'd love to see what you all have seen," Brad said pensively.

Even through the shielded walls, they heard another blast of thunder, and with this one, the lights dropped to black. As they flared up again, the monitors flashed red alarms.

"That's it!" Xavier thundered to match the weather. He left the room, calling over his shoulder, "Kat—a little help?"

She reached up and touched Brad's elbow. He could feel the lava in his veins.

"Sorry about all this, Brad. The entire building is supposed to be protected from this sort of thing. Power surges are quite unhealthy for most of our experiments. Clearly, something is not set right. We'll be back soon."

"No worries." Brad took a sip of coffee, feigning indifference. "I'll just hang tight here. Take your time."

As Katya opened the door, Bradley could hear Xavier chewing out some poor tech for "not securing surge protection during a storm," and something about a Faraday cage. He decided to make the most of the opportunity and took a look around. One of the time pods was closed, but the other lay open for all to see. Brad climbed several steps to its platform and peered in.

"Not too roomy, but I guess they do the trick." He tried to feel the interior, but he couldn't quite reach. Leaning over, he put his knee up on the rim, bent forward, and patted the side of the pod. It was padded, soft. Another alarm sounded, and this time, the lights died completely. Emergency lights flared, painting the room in shades of crimson. Brad tried to straighten up, away from the pod, but his new shoes proved his undoing. He slipped on the upper step. His coffee sloshed, and in his attempt to keep from making a complete mess, he felt himself lose his footing. His knee slipped over the pod's edge. He fell into the pod and rolled. The top slammed on his foot. He pulled it inside instinctively and rubbed it as the lid slid his shoe off and then slammed shut. He was going to make quite the impression on Kat and her colleagues.

* * *

The lights came on again, brightened, and then steadied. Katya was the first back into room. Xavier followed close behind. The room was strangely quiet, except for a faint, chiming alarm at one computer station. Both pods stood empty. Beside pod number one, a puddle of coffee dribbled over the edge of the top step. Next to it lay one of Bradley Glenn's new shoes.

49

Scrambled Screens, Second Thoughts

"Mr. Dresden."

Dresden knew the voice immediately. It chilled him. "Yes, Mr. Ambassador?"

"I've received some of your P.I.'s files. My engineers tell me they are surprisingly helpful." Dresden let out a long, quiet breath as the man continued. "As soon as he returns and you are in possession of the drive with the rest, please terminate his employment."

"Yes, sir," Dresden said. "With pleasure."

"Even now, we are in the process of changing some of our hardware and software. I expect progress to be going forward at a breakneck speed. Please stand ready for our new Latin-speaking recruits."

"I'm ready," Dresden lied. How could one prepare for the onslaught of an ancient army? He wondered briefly if they liked spaghetti. Or pizza. How many pizzas would it take to feed a Roman legion? He scratched his chin and looked out the window. The lights of Fort Collins were coming on like twinkling stars, a waning Moon just rising in the east. He removed his cowboy hat, placed it carefully on his desk, and wondered why Rex hadn't returned his call.

It was time to begin planning an exit strategy.

* * *

The coffee on the floor alarmed Kat. Any liquid around all this electrical equipment was dangerous.

"No," Xavier gasped. He dashed to a screen. It was completely scrambled, but the other one told the tale. It displayed numbers indicating that both time pods had been triggered.

© The Author(s), under exclusive license to Springer Nature Switzerland AG 2021
M. Carroll, *Plato's Labyrinth*, Science and Fiction,
https://doi.org/10.1007/978-3-030-91709-8_49

"Better get Todd," Kat said, grabbing a company terminal.

Xavier leaned over to Pod number one and picked up the shoe. "Recognize this?"

"Great," she said. "Just great. That must be Brad's coffee all over the floor."

Todd entered and made a beeline to the monitor. "That's not all. Both pods transported someone. The storm knocked out our CCTV in several rooms, so we can't see who hopped in."

Kat felt the heat burn its way up her neck. "Who could have been in the other one?"

Brianne burst through the door. "I've just alerted security. We can't find the inspector and he never checked out."

"Security won't find him," Kat said.

Brianne stared at the empty pod. "You're kidding."

"Nope."

"I should have kept closer tabs on Rex." The new recruit looked sheepishly at the floor.

"Rex?" Xavier and Kat barked in unison.

"The idiot inspector," she clarified.

Kat felt the terror building inside as she studied the screen. "It's all set for Thera. Same settings we used last time."

Brianne began to pace. "This is bad. This is very bad. The man is even dumber than I thought. Why would he wander into the arena?"

"I doubt that he just wandered," Xavier offered. "Something's up."

"Boss, I'm telling you, the man is incompetent enough to get lost in his own apartment."

"We've only got two pods," Kat said. "How can I go back and rescue two people?"

"Don't panic," Todd said. "I've given that some thought. Remember that those pods simply disappear in the other timeline, and they're always present here. You can go back with three talismans, return those two here, and trigger your talisman to call another pod for you. In effect, when Xavier came back for me in France we did a dry run of what you'll be doing."

"Will they be okay?" Brianne asked.

"I can only imagine how surprised they are," Todd said. "I'll go back with Katya."

Xavier had that unstable look again. In fact, Kat thought he looked as if he was losing touch with reality. Gone was the encouraging leader. This boss looked like he was about to come unhinged, and it worried her no end. Xavier shivered. "No, no. It's too risky. We've never had four out before. Three is enough variables."

"Yes, I agree," Katya said.

"You would," Todd argued. "Then let me go alone."

Katya stepped over to face Todd. She put her hands on his shoulders. "Todd, I've got this. I need you here to watch the engineering side, in case we can't get back."

Todd looked at Xavier and Brianne, but they merely nodded agreement.

"I really hate that," he snapped.

"Hate what?"

"I hate how much sense you always make."

Xavier said, "Let's get you ready. We need to assemble a full team."

"I can have all hands on deck by five, if they haven't left yet," Katya said.

"I'd say this qualifies as an emergency. Brianne, call in anybody we need. Katya, go get the world's fastest snack—you'll need the calories—and we'll shoot for a 7 pm departure."

Katya headed for the lounge, but the unknowns haunted her. Would the currents of time take her somewhere else? Who knew what aberrations the electrical surge introduced to the system? Assuming she made it back to the right spot in time and space, would she be able to find Brad and the inspector Rex?

All she could do was close her eyes, jump in, and hope for the best.

* * *

Katya was wolfing food in the lounge when Xavier appeared, wearing an apologetic mien. He had Ajit in tow. "Look who I found pacing the lobby," Xavier said. Ajit gave her a hug, and they both sat at the little table as Katya continued to pick at her food.

Ajit Joshi had an acute sense of time, his daughter reflected. He knew all the nuances of the past, the ebb and flow of the fossil record. He could mentally break the Paleozoic Era into its components: Cambrian, Ordovician, Silurian, Devonian and so on, and each of those he could dissect into its own smaller segments. Yes, Ajit Joshi had a fine sense of time. But not of timing. He usually got things wrong. He was early or late or in the wrong place or the wrong time, as he was now, at ChronoCorp. The place was in crisis mode.

"So, I understand you are having some technical challenges," Ajit said. "I won't stay. I just wanted to drop by and…I understand you've had some power surges. Any damage?"

Katya put down her fork and leaned toward him. "Dad, it's more serious than that."

Ajit looked up at Xavier, whose concerned expression matched Katya's. "Oh? What is it?"

She put her hands together on the table and laced her fingers. "It's Brad. He's missing." She felt the burn in her eyes and realized that, despite his stubbornness, she had a soft spot for the man. Brad and his puppies and kittens.

"I'll just—" Xavier said, not finishing his excuse. He left the room.

Katya looked down at her unfinished dinner. She put her fork down and stood. "Walk with me—I gotta get to the Arena. We've got another person missing, too. Call him an innocent bystander."

"I suppose you're finally having second thoughts about all this?" Ajit asked, following her down the corridor.

"Whatever happened to changing the course of time?" she said over her shoulder, picking up her pace. "All that stuff about saving zillions of people by putting in a good word here and there to scientists of old? Was it just hot air?"

"Yes—I was caught up in my emotions. But I always knew that this thing was fraught with dangers you couldn't control."

"What about the people we could save? And Mom?"

The answer Ajit gave surprised her.

"My dear, the essence of people, their influence on the world around them, ebbs and flows. Do you think your mother's loss had only negative consequences? I was devastated, but it forced me to make changes. All those changes in attitude, in interest and focus, led to my new work, my awards, our discoveries. Those discoveries in turn gave birth to more unpredictable events, many of them breakthroughs that have changed the course of paleontology and biology, and they will continue to do so. The vacuum your mother left has been filled by those she mentored, and they're taking her research in new directions. Her own student was one of the Nobel finalists last year, did you know? Your mother's death was a tragedy, yes, absolutely. I feel it every day. But for one person—or a corporation—to decide whether someone lives or dies, or whether someone will have knowledge that did not come to them in the natural unfolding of time, that's way above anyone's pay grade."

"So you think these questions are just too big for us puny little humans?" Katya flared. "That doesn't sound like the scientist-father I know. It's when we have pursued the big questions, when we have looked the terrifying things of life in the face, that greatness happens."

"Greatness? By whose definition? And at what cost? You've got people getting lost and maybe dying. And others stealing your stuff; corporate espionage. What will they be doing with it? Building a superweapon? Assassinating someone in the past?"

"It can't work that way," she shot back.

"So you've said, but it seems to have that potential. No, Kat, this whole thing is getting away from you, and as far as I'm concerned, you're the good guys. What's going to happen when bad guys can build their own version?"

She threw her napkin down on her paper plate. "A debate for another time. Right now I have to try to find Brad."

"Yes, do try."

50

Lost in the Fields of Time

Bradley Glenn never got headaches, but he had one now. A doozie. It came with a free side of vertigo. Coffee soaked his shirt, pants, and the interior of the casket he'd fallen into. He fumbled with the hatch. He didn't realize something had changed until he swung the top up. The blue light of the outdoors flooded in. Scents of evergreens, smoke, and sun-baked stone washed over him like the coffee had done moments ago. He heard the call of birds and the wind in the trees. Had the ceiling been blown off of ChronoCorp in one of Colorado's famous tornadoes?

He sat up, saw spots, and lay back down, still cradling his half-empty cup of coffee. As soon as the world stopped spinning, he sat up again and tried to get his bearings. The walls of ChronoCorp had transformed themselves into a grove of trees. Purplish fruit hung from the branches. Their twisted trunks hid the landscape beyond. The sunlight was warm, almost hot, but his blood ran cold.

Climbing from the pod, he felt the sharp sting of gravel on his bare foot. Carefully, he picked his way down a steep, rocky slope, thought better of it, and made his way back uphill for a better viewpoint. Past the pod, the grove cleared, revealing a zigzag path leading to a high temple of some sort. Stylized horns crowned the building's corners, and smoke wafted from the interior. Smooth, polished stone covered the walls. The structure perched on the crest of a rocky outlook, its doorway framed by red columns. Brad turned to look back toward the grove. On the steep slopes below, among vineyards on cultivated terraces, brightly colored buildings scattered within the greens of pine forests. There were no cell towers, no electric lines, no airplanes in the air or paved roads on the ground. The sounds of modern life, car engines and horns,

© The Author(s), under exclusive license to Springer Nature Switzerland AG 2021
M. Carroll, *Plato's Labyrinth*, Science and Fiction,
https://doi.org/10.1007/978-3-030-91709-8_50

phones ringing, trains and subways, had been replaced by birdsong, the bleating of sheep and distant voices in a quiet world. It was beautiful. It was petrifying.

He stepped onto the pathway. It led uphill toward some voices in the temple, and downhill toward a patchwork of farmlands. As he tried to decide his course of action, he caught a few words from the conversation above. They were foreign to him, so he decided to avoid contact until he absolutely needed help. He turned and began a descent back toward the shelter of the trees. But what was to the left? Low-lying buildings surrounded a sort of courtyard. He off-roaded to the little grouping of structures. He crept down an alley between several two-story buildings. He was no architecture expert, but he had never seen such designs. Their stuccoed walls had irregular window openings, and the sounds of conversation came from the second floors. So did the smell of cooking food. These must have been homes. At the center of a small tiled piazza stood several sculptures and a fountain. The water was steaming and smelled of sulfur. He remained at the edge of the alley, studying the elegant mosaic floor. Its colorful patterns reminded him of Greek pottery he had studied in art history. Greek. And that had been the subject of the press conference. Could it be? Suddenly, he felt very exposed.

Boisterous laughter came from overhead. There were people on the roofs. He decided this was no place to tarry; the peril of getting caught outweighed the excitement of his discoveries. Brad retraced his steps up the hill, limped off the road and ducked through several branches toward the denser forest. A voice rang out.

"Hey—hey you! Boy am I glad to see you." A man wearing some kind of uniform approached him unsteadily. "American?"

Brad nodded, shocked into silence.

"I don't know what the f—what the *freak's* going on. One minute I'm at ChronoCorp, and the next?" He waved his hand at their unfamiliar surroundings. "So what is this place, anyway?"

"Haven't got a clue," Brad said. "Greece, maybe. Are you okay?"

"Okay? I wouldn't say that. I just instantly ended up somewhere that I wasn't. How is that okay?"

"Hey, don't worry. We'll get through this. You and I. Together."

"Sounds a little like whistling in the dark."

Brad shrugged. As he did, his shirt tore at the shoulder.

"Man, who's your tailor?" the man laughed, but as he did so, one of his pant legs came loose.

What was left of Brad's coffee began to squirt from several pinholes in the cup. He put it gently on the ground, not taking his eyes off of the stranger. He stood and said, "So what's your story? Do you work for them?"

"For ChronoCorp? Oh, no. I'm a state inspector. I was there to examine the pipes." Brad noticed the uncertain tone to the man's voice.

Brad stuck out his hand. "I'm Brad, and I don't work for ChronoCorp either."

The inspector relaxed visibly with Brad's comment and shook his hand. "Rex. Who do you work for, anyway?"

"I'm a paleontologist at the American Museum of Nature and Science in New York. I flew out. They gave me a tour, but I think that thunderstorm scrambled their controls or something."

"The thunderstorm. That explains it." He nodded knowingly. "Paleontologist, eh? So you study…different diets? Paleo? Vegan?"

"Ancient life. Dinosaurs. Mammoths."

"Ah, I see. Sure." Rex's belt disintegrated. Brad's remaining shoe came undone at the seams.

"Geez," Rex squealed. "What's with our clothes?"

"It's more than that," Brad said, pointing to the coffee cup. It looked like someone had taken a blowtorch to it. Rex's old-school clipboard splintered into several pieces. He seemed to be hiding something under the flap of his coat. He fumbled and dropped it into the gravel at his feet. Brad picked it up and recognized it immediately. He handed it to Rex.

"So you inspectors must need to take a lot of notes. This is quite the little drive. What is it, a terabyte?"

"Something like that," Rex said, stuffing it into what was left of his pocket. It fell through and turned to black and chrome powder on the ground.

Brad raised an eyebrow. "I'm not very well versed in technical stuff."

"This is ridiculous," Rex whined. "We gotta get back."

"It seems to me our best bet is to stay close to those contraptions we arrived in."

Rex looked up the hill, and then back at the grove. "Yeah, maybe so. Maybe they've got a 'return to sender' button or something."

They crossed through the barrier of trees and back into the clearing. Two rectangular indentations flattened the grass. The pods were gone.

* * *

Katya's father was right, of course. Things had gotten out of hand. Now, Brad was gone. And it was her fault. She was the one who insisted on inviting him here. She was the one, ultimately, who left him on his own in a room full of dangerous equipment. She fought the sting in her eyes, ran into the restroom, and sat in a stall with a wad of toilet paper. She let the thoughts blow through in ragged sobs. *Consequences.* She had been so cavalier about this fearsome power ChronoCorp now wielded. But no more. People like Bradley Glenn were far more precious than a few data-filled scientific papers. He had a gentle air about him, that kittens and puppies thing, combined with some sort of inner strength. Or maybe she was just remembering some description of a rom-com celebrity. She was definitely confused and in no shape to be gallivanting across ancient Greece. But she had made her bed.

So now to the rescue mission. Brad and the government guy could be anywhere. Had they headed down to the coast? Gone up the hill and been captured by someone? What if they had split up? How much of Minoan time had elapsed since they left? She might arrive a week after they did. Or a month. And with the elapsed time, their geographic point might have drifted. Were they even on Thera?

She took in a slow breath, dabbed her eyes, and pushed back through the door. There was work to be done, too much to do by herself. She prepared to make the case to Xavier that she bring Todd along. But before she could, Xavier interrupted.

"Kat, there's just too much territory to cover. I think it's best if Todd accompanies you on the search. I'll take the comm."

Katya felt a deep sense of relief. Now, she thought, they'd have a fighting chance.

"I'm worried about that quantum drift." Todd sounded urgent. In fact, he seemed uncharacteristically nervous, his talisman swinging from his neck. "We'd better get going."

Emil came through the door carrying two of the newest version of his platypus polaroids. He handed one to Todd and one to Katya. "Here are the little beasties, ready to record."

Todd nearly dropped his. Katya noticed his hand twitch as he grabbed the pulsing blob. He looked at her, beamed almost maniacally, and climbed into his pod. She had second thoughts about travelling with him. Why was he so strung out? But there was nothing for it. She grabbed the other three talismans, climbed into her pod and waited for Xavier's command to transport.

*　*　*

Xavier peered at the pods. Todd and Katya had left the building. His eyes darted around the room. The walls seemed to be closing in. Someone had sucked all the air from the Arena. He could feel the sweat rivering down his back, and he just couldn't catch a breath. *What if they fail? What happens when someone dies in the wrong time?*

Now he was dizzy, nauseated. He couldn't stay in the Transport Arena for another second. He tapped his headset and steadied his breathing. "Emil, come down and take the comm, will you?"

"Sure thing, boss."

"It might be awhile."

"Be right there."

Xavier was about to crawl out of his skin. He stepped into the corridor. The air was cooler outside of the cramped Arena. Emil appeared and Xavier cornered him. "Emil, I need you to crunch some numbers for me."

"You know how much I love to crunch."

"I do indeed. So here's the idea: could a closed time-like curve somehow twist back on itself, so that you might have duplications of something?"

"Duplications? Twinning of events?"

"Or even people?"

"That's got to violate all sorts of logical flows. Paradoxes galore."

Xavier's breathing was becoming labored. "That's certainly what it sounds like, but what does the math tell us? Is it possible?"

"Might be. I doubt it, but it might be. Maybe in a parallel universe close to ours? But I don't have a clue how we would access another universe like that."

Xavier grimaced. "I think I do. We've been traveling back in time, but we never went the other direction. Forward."

"Into the future," Emil muttered.

"Just a little bit. Perhaps a few milliseconds, but to the same place in 3D space. That might end us up in another universe, but in the same 'place', yes?"

Emil seemed to be warming to the idea. "I'll look into it." He gave Xavier a sideways glance. "If we only displace ourselves by a couple milliseconds, the two universes might simply fade back to where they were. Reset."

"But there might be some time before they did," Xavier offered.

"It's certainly not something I would recommend. You're not… You're not entertaining thoughts of trying something like that, are you?"

"We're researchers. We experiment."

"Boss, you look like you need a break."

"That's exactly why I had you come down."

Emil said, "Look, my equations are quite theoretical. I wouldn't stake my life on them. We need much more work on this sort of thing."

"But of course. No worries. You know me." Xavier made a mad dash down the hall, into the elevator, and outside. He inhaled the bracing Colorado air, let it calm him. But he felt the old feelings again, the ones he'd endured after all of his recent slips. They were still there, simmering away. He began to tremble.

Perhaps it was time to take matters into his own hands. ChronoCorp's accidental discovery of time travel was becoming a curse, not the blessing he had hoped. But he could use its own technology to undo the damage, to remake the universe into its former nature. *Yes,* he thought. *Maybe it's time to do some quantum cleanup.* But first, he had to call his doctor. These panic attacks were getting out of hand.

51

Reunions

"What gives?" Rex said. "Did someone steal our...our tanning booths?"

"No, they were too heavy. I think they dissolved...or something."

Rex reached to an overhead branch and pulled off some of the purple fruit. "Right. Into thin air?"

"Into time. Rex, don't you know what's going on here?"

Rex shoved one of the fruits toward Brad. "Fig?" Brad took one. Rex said, "Why don't you enlighten me."

"What's the last thing you remember at ChronoCorp?"

"I remember closing my eyes and praying that you and the ChronoCorp corps didn't find me. See, I wasn't really supposed to be in that room. It was an accident, really. I made a wrong turn and..." His voice faded off. "Just what are you getting at, dinosaur man?"

The laughter of young girls drifted through the trees. Brad leapt to the edge of the grove. Coming down the path, a large group of teenagers carried baskets. He spoke out of the corner of his mouth. "I think they're coming this way. Fig-picking expedition."

"We gotta get out of here!"

"Yeah, Rex, I think you're right."

* * *

Katya opened the pod hatch gingerly, feeling the throbbing in her temples. She sat up slowly, letting her vision clear. She was worried about Todd; he had taken so many more Q-slips than she had. But he was already standing up,

M. Carroll, *Plato's Labyrinth*, Science and Fiction,
https://doi.org/10.1007/978-3-030-91709-8_51

leaning with his back against his pod, arms across his chest. He wore a cavalier grin.

"Guess what?" he called. "No headache!" He lowered his voice conspiratorially. "I took something. Several somethings. Worked like a charm."

Katya swung her legs over the side and stood. Her main concern was to remain upright. Her second was to find Brad. It would be nice to find the poor lost government guy, too, but she had to admit that he was a distant third on her list. "Good for you. Any sign of them?"

Todd came to her side, walking in serpentine fashion. Katya had never seen him drunk, but he was doing a great impression of inebriation now. "Hey, we'll find them," he said. "Not to worry, my esteemed colleague. Let's get a move-on."

"I think we should stick together for a little while. If we can't find them, we can split up later."

"Sure, you're the boss. Hey, wait a minute. *You're* not the boss. Neither am I. Guess we can do whatever we want."

Their arrival point had drifted considerably uphill of where they had materialized before. They couldn't see the olive grove where they had arrived on the last Q-slips; this landing site was more exposed. A path meandered up the mountain. In the distance, on a high promontory, stood one of the Minoan peak sanctuaries. Katya led the way in that direction.

After a few minutes of hiking, Todd held his hand to his mouth and yelled, "Bradley Glenn!"

Katya rounded on him. "Are you nuts? We need to be clandestine about this. We've got to get those guys out of here with as little fanfare as possible."

Todd grinned and shrugged. Katya realized that he was, in fact, high. She leaned closer, looking into his dilated eyes. "Just what did you take?"

"A bit of this, a bit of that."

"Can you be more specific? It's for science."

He pointed at her. "You're funny. I took a dose of hydrocodone for the pain and a couple hits of meclizine for nausea. The meclizine is over-the-counter stuff. Satisfied?"

"How much?"

"I stuck to the prescribed amount of the opiate. Stuff's dangerous. I took six meclizine."

"How many are you supposed to take? Normally?"

"Up to two. But there's nothing normal about this. I'm good, though! I can even touch my nose with my finger. Want to see?"

"Not necessary. Let's just find our guys."

A distant rumbling told them that Thera was restless. The ridge obscured the bay from which it rose, but it was there. Katya stopped and looked at a walled garden. The garden had been well tended, its vines and trees mature. The stonework sent a chill across her shoulders. She spun around and scanned the landscape, then turned back to the wall. "Todd, how long would you say that wall's been here?"

He shrugged. "Awhile. It's got grass growing out of it. The stones seem like they've been settled here for a long time."

He was right. Green blades, sharp as sabers, rose between the squared rocks, leaning against the gentle breeze. "Years?" she asked.

"Maybe. Why?"

"That rocky buttress over there, I remember it. It was at the edge of the site where we arrived last time. And this wall and this garden *weren't here*."

"I wonder how much time has passed. We arrived at a slightly different place and a different time." The volcano bellowed again.

They continued their ascent, but in moments, the branches of a nearby hedgerow snapped. An explosion of leaves scattered and Bradley Glenn exited the foliage, with Rex close behind.

"Kat!" Brad yelled. "Well, if it ain't my favorite physicist."

Katya had always scoffed at those melodramatic love songs talking about the desert missing the rain, or the moon longing for the stars. But she was feeling a bit melodramatic about now, enough that her eyes stung. She hadn't realized how much she missed him, or how worried she had been. It wasn't the professional or practical need of recovering him from the time travel fiasco; she had actually grown attached to him. There was no question of protocol now; he threw his arms around her and gave her a joyful bear hug.

"Todd!" Rex called out. He lunged for Todd and gave him a hug, too. Todd's reaction was considerably different from Katya's.

"Glad you guys said something," Brad said, glancing behind himself. "We had no idea where to go. I was hoping somebody would show up to clear all this up. But we've got a gaggle of teenaged girls on our heels. They'll be here any moment."

"Shall we get you back into your pods?" she asked Brad.

"Funny thing: they're gone."

Katya looked toward the bushes, and then looked down at her tattered clothing. "We need to get back immediately. We're parked down there."

The foursome scampered downhill to Kat and Todd's point of arrival.

As they rounded a small boulder, Katya gasped. "I don't understand. We haven't been gone that long."

The pods had vanished. Todd knelt on the rocky soil and pointed at a shallow indentation. "This is where they used to be all right."

Katya held up the talismans. "We've got four of these. Let's hope they all do what they're supposed to." She passed them around with a word of warning: "Don't do anything until I tell you to."

Rex looked at his lanyard, askance. "Don't worry; don't worry."

"Todd, how about if you go first? Pave the way for the amateurs. I'll bring up the rear."

"Sure." He sounded tentative.

"Just push the button and we'll see what happens."

He did. In moments, they felt a whoosh of air as a pod appeared. He stepped in and lay down. Katya closed the hatch. Todd pressed the talisman again and waved as the pod faded away.

"Nice!" Brad said.

Rex flinched and blinked at Katya. "That's what was supposed to happen, right?"

"Yes, absolutely. Now, Rex, hit your talisman."

"Like this?" A second pod materialized.

"Good," Katya said. "Now, just step in and close the lid, and then hit that button again. Hopefully, this is your ticket home."

As Rex leaned over the pod, he asked, "Hey, just where are we, anyway?"

"Just get in. We'll have plenty of time to chat later."

Rex rolled in to the new pod with the grace of a beached whale. Brad closed the hatch. Brad and Katya stepped back. A bright flash announced the pod's exodus.

"Your turn," she said.

"Kat, I'm sorry about all this mess. I feel like it's my fault."

She smiled and patted his shoulder. "Don't worry. You haven't had much practice traveling through other dimensions. I'm sure it will get easier."

"Most things do. With practice. Rex was pretty nervous when we first got here. I had to sort of talk him down."

"Puppies and kittens and now inspectors. Thanks for that."

Brad looked up the hill in the direction he had been exploring. "I was wondering the same thing, though, about where we are. I saw some patterns that looked Greek."

"We'll get to that." She pointed at his talisman.

"Hey, I saw the press conference. Thera, right?" His voice began to ramp up in pitch. "Big volcano? Murals and palaces? A very long time ago?"

"You're not just a pretty face. Now, let's do this." She pointed at his talisman.

"Wait. Just wait a minute." He grabbed her shoulders gently, then dropped one hand from her, slowly. She felt his hand brush across the muscles she'd built up in her years of kickboxing, and wondered if the firmness would turn him off.

"You're always in such a hurry," he said. "You toil on your monitor all day, but all this time, those numbers that you come up with result in—" He swept his arm around the view. "This. You should give yourself a chance to see what wonders have come from your labors. Give yourself a moment. We could stay here for awhile. I found a beautiful mosaic I could show you next to one of those stinky fountains."

Kat looked up at a column of smoke in the sky, the mark of an angry volcano. "I have bad news. This may not be *when* we thought it was. We may be six years later than the last trip. Quantum drift. This place may blow sky high at any moment. On the other hand, we know most people got out before the explosion, and the place seems fairly well occupied." She looked longingly up the hill.

"It's just up here, across a plaza," he urged.

By the time they reached the little enclave, Katya's clothing was in shreds, and Brad was nearly naked. They both tried to avert their vision; the whole encounter was starting to feel a little too much like Adam and Eve in the Garden of Eden. They were relieved to know that the "robes of return" awaited them soon. Keeping their eyes turned away from one another, they took a moment to enjoy the little courtyard.

Katya noticed Brad had gotten quiet. Was it sorrow? Regret? She resisted the urge to turn and study him. "Hey, it's beautiful," she encouraged, "just like you said. I'm glad you kidnapped me, at least for a moment. Good for me, right?"

"It is, yes. It does make me wonder, though: is that all there is to civilization? A slow rise to glory and beauty and great advances, only to end in a pile of rubble?" The ground muttered beneath their feet.

Katya was silent. He gazed at the little village before them. "Look at the copper rooftops, the tiled alleyways. And the fountains? Stunning! The builders of this place are probably no different from you and me, Kat. Once upon a time, this place had an everlasting future, a future that ended before its time. Did its inhabitants have lofty goals? Were they fulfilled before their culture was snuffed out? Now all those palaces and plazas are gone. And we're left in our own time, three thousand years from this moment, to reassemble all the little crumbs and scraps."

She tilted her head. "And isn't that what you're faced with every day? Picking through the bones and footprints and skin impressions, putting the puzzle back together? The dinosaurs had their time in the sun, just as the Minoans did."

"But the beasts I study didn't have the brains to do anything about it. You'd think we would be smart enough to learn from our past."

A statue toppled from a rooftop as another tremor hit. Smoke billowed from several spots on the forested slopes above. She brushed the tip of his nose with her finger and turned him downhill. "We've got to get you out of here." At the top of the mountain, a loud crack echoed like the shot of a cannon. House-sized boulders sailed into the sky. "We've got to get *us* out of here."

They rushed back to the lower grove. Brad hit his talisman, and another pod materialized. He flashed his teeth and climbed in. "Kat, don't be long. We'll be waiting. See you on the other side, whenever that is."

In moments, the pod was gone.

"My turn," Katya muttered. She held up her talisman like a magician with her magic wand. She frowned. She hit it again. Nothing happened.

52

Makeover

Emil's quantum calculations had come to a nebulous conclusion about Xavier's question. But some of the math sounded promising to him. Xavier had returned to join him after only a few minutes. Emil was grateful. He had no appetite for running a Q-slip on his own.

While they awaited the return of the other travelers, Emil called Xavier to a monitor displaying his equations. "See this expression here? What you get looks like a closed time-like loop that flips into a Mobius strip, a figure eight. But you would have to be careful. If this Mobius time curve thing really played out, a person could actually end up with a parallel doppelganger in the same universe. Two Xaviers running around. There's a thought."

There was a thought, indeed. *How do you rewrite a history that has already become so complex with people and events?* Go back to the beginning, of course. But how far back is that? Back to the beginning of ChronoCorp? Back to the invention of the ring laser? Back to Einstein? If Xavier assassinated Albert Einstein—a thought he would have found repulsive only a few weeks ago—the world would suffer so. The scientific loss would be unimaginable. Perhaps he should choose someone later, someone less influential, but still concerned with the mathematics or hardware of time travel. Was that any less repulsive? It was as if he had become a different person in the last few weeks. Or perhaps he was simply seeing reality more clearly.

At that moment, it became clear: Arvin Devereaux. Devereaux's modern computations had transformed ChronoCorp from a quantum physics research lab to a time portal. Briefly, Xavier wondered what kind of family Devereaux had, who his friends were, whether he had grandkids. He was an old geezer, certainly old enough to have grandkids. Maybe even great grandkids.

M. Carroll, *Plato's Labyrinth*, Science and Fiction, https://doi.org/10.1007/978-3-030-91709-8_52

Devereaux had lived a good life. Bringing an end to it would be no great loss to him or the world (although Xavier supposed he would have to go back to a time earlier than most of those things came to pass). Yes, these were distractions, things that might derail Xavier from his mission. He had to focus, for the good of the cosmos he knew. And once he took care of ChronoCorp, he would need to do something about the Primus Imperium.

But Devereaux might not be enough. ChronoCorp had lots of really smart people. Those designers and engineers and theoreticians just might stumble onto the time travel thing, given enough time and resources. So he had to undo the past on another front: the resources. Before Mila van Dijk came along, the time travel discoveries were still largely theory. It was her backing that gave the corporation a boost and a change in direction. No coins, no ChronoCorp. That came to two assassinations he would need to do. It was a shame: he liked Mila. She was spunky and sharp.

He would need to do the deeds at close range, because he only had his handgun. So be it. There should be nothing impersonal about what he was about to do. He owed his victims at least that.

Emil broke in on his thoughts. "We've got another incoming pod."

Brad stepped into the lab. "You may want to hear this."

*　*　*

Chelsey Grosvenor eyed Allan McElroy. "Nervous?" she asked.

"I just keep thinking of Fred and Joslyn."

"And don't forget poor little Alfie. But we know what went wrong, and thanks to our sources, we know what needs to be done. I have confidence that our techs have their ducks in a row."

The man lowered his eyes to a small console. "And I have confidence that our dear Ambassador wants a win badly enough to sacrifice a few more people." He wiped his brow and tapped his touchscreen. The cabin was cramped, made smaller by his edginess.

Outside of the temporal capsule, engineers and technicians scampered over equipment and monitored screens. Through the glass, Grosvenor and McElroy could hear the master countdown. The numbers droned on, and when they reached thirty, all noncritical personnel vacated the Transference Theater of the Primus Imperium.

"I guess this is it," Grosvenor said. "Orville and Wilbur. Norgay and Hillary. You and me."

"Laurel and Hardy," he added.

Chelsey reached over to shake McElroy's hand. His was slick with sweat. "We'll be fine," she told him confidently. They both braced themselves and looked forward, toward the past.

*   *   *

The Ambassador turned away from the busy Transference Theater and addressed Dresden. His voice lacked the poison it had in their last meeting. The Ambassador had adopted a business-as-usual demeanor once again. Dresden wondered if the man even remembered their last meeting. "We must prepare for the onslaught. When McElroy and Grosvenor return—not if, but when—I will want a follow-up test within two days. And this time, we'll take the troop transport with full crew. If all goes well, we will be welcoming crack Roman soldiers by this time next week."

Dresden nodded toward the Theater. "Where are they going?"

"Nothing too ambitious. The Loire River Valley of the nineteenth century. It's close to our ultimate destination in the three dimensional plane, but not too far back in time. Very conservative."

Dresden tried to work out what was conservative about tossing two people a century into the past and having them end up on the other side of the world. "Wise strategy," he told the Ambassador.

A klaxon blared, announcing the return of the travelers. One of the technicians called out, "Power levels at threshold. Five seven."

The monitors dimmed and flickered. A flash of light flooded the dim Theater, and the temporal capsule materialized. Both the Ambassador and Dresden focused on the door as the technicians approached. But before the techs could get there, the hatch popped open to the grinning faces of Grosvenor and McElroy. The techs ushered them to the Ambassador's viewing balcony.

"Well?" he said. "How was it?"

McElroy said, "Glorious!"

Grosvenor added, "Especially if you like vineyards."

53

Choosing Sides

When Rex Berringer arrived at ChronoCorp, even after a trip of 3,500 years, he could tell something was amiss. Xavier and Brad stood across the Transport Arena, glaring at him as he stepped unsteadily from his pod. Xavier leaned with his back against the wall, his arms across his chest. Brad looked a little more tentative. Another man stood nearby, with a straw hat and a lot of dust on him.

Rex regained his balance, straightened, and chirped, "What a trip! That was amazing."

Brad's and Xavier's expressions didn't change. Xavier held up his index finger, pointing to the heavens. "I can't for the life of me understand why a state plumbing inspector would need a robust solid-state flash drive to check some pipes. Care to illuminate me?"

"And while you're at it," Brad interjected, "you might want to clear up a mystery that their security people have. There doesn't seem to be an inspection scheduled for today by anybody, and the county doesn't have an inspector named Rex *or* Kelsey Acres."

"Oh, the name. See, when I'm at work I go by…" Rex's voice trailed off as he realized how disinterested they were. The jig was up, and his most recent employer could not save him now. Anyway, with all the cloak-and-dagger threats from Dresden, Rex wondered whether his employer would even want to.

"Look, I'm a private investigator. My name is Rex Berringer."

"At last," Xavier said. "Some refreshing frankness. Go on, please."

"I was supposed to get the ChronoCorp files—as many as I could, anyway—pertaining to time travel."

© The Author(s), under exclusive license to Springer Nature Switzerland AG 2021
M. Carroll, *Plato's Labyrinth*, Science and Fiction,
https://doi.org/10.1007/978-3-030-91709-8_53

A look of trepidation crossed Xavier's face. "For whom?"

"Most of what I got was on that drive," Rex went on, "and it's turned to dust back there, wherever we were."

"But not all of it?"

"I managed to get some contingency files out remotely. You guys have some great security, by the way. It was frustrating as hell." He could tell that Xavier was doing a slow burn, so he tried distraction. "I can tell you more when our rescuer gets back."

His mention of Katya proved to be the perfect diversion, because Xavier began to chatter with the engineers. The question on everyone's mind apparently was: where was she?

*   *   *

"Okay," Katya said evenly. "Okay," she said again, with no one there to hear. "Just what is it that's okay about this situation?" she blurted to the gray-green sky.

Perhaps she needed to go up to where the other pods had been, where Brad and Rex landed. But that seemed silly. They had traveled all this way. Why would a few feet matter? She was grasping for straws, but there was nothing left to do but grasp.

She was on her way up the hill when she heard the fourth pod materialize. Through the last branches she could see the welcoming glitter of sunlight on the chrome object. "Guess these things are a little hard to predict," she muttered. She lunged for the pod and departed to the sounds of adolescent fig-pickers.

*   *   *

Katya arrived just in time to catch Todd massaging his temples. She leaned over to him and whispered, "Are you getting what Xavier has?"

He rubbed his eyes. "Maybe."

"Maybe those drugs aren't such a good idea?"

Todd just groaned.

Katya spotted her father in the corner. He smiled at her and she grinned back. Then she noticed the group of engineers, including Xavier and Brad, encircling Rex. In fact, they seemed to be corralling him in. Brad left the group, wrapped in the traditional "robe of return," rushed to her side and stopped short. "Welcome home."

She eyed Rex. "What's up?"

Brad turned to pat Rex's shoulder. "It seems our inspector here was borrowing ChronoCorp files."

"So the state of Colorado…." Kat realized her mistake as she took in Rex's sheepish grin. "Oh."

"Look, look," Rex said, waving his hands toward the floor. "I'm happy to tell you all the details…for a price."

"It seems you're in no position to barter," Xavier snapped.

"We can sue him, have him locked up," Brad suggested.

"Yes," Rex said. "You do that. And in the meantime, I'll have a visit with my employers and tell them all about your work here. In more detail than they've got already. That could cause a lot of heartache for your dear ChronoCorp."

"Nonsense!" Todd crowed, but Xavier held him back.

"No, he's right. We can't afford that kind of brain drain with a potential competitor." Xavier frowned and thought for a moment. Finally, he said, "Okay, Rex, if we agree to not press charges, will you tell us everything you know about your client?"

"After all my trouble, I'll need a little spending money. I'll need to move to someplace…warmer. I like palm trees."

"I'll send you someplace warmer all right," Todd snarled.

Rex faced Xavier. "Take it or leave it. Do we have a deal?"

"Spill it, pipe boy," Brad said. Katya smiled.

"I'm afraid it's bad news," Rex said with a sarcastic edge. "Those files aren't the worst of your troubles. I was supposed to deliver them to the head of the Primus Imperium. They call him the Ambassador, and I'm pretty sure he's into a whole lot of dark crap: arms deals, maybe drug cartels, I don't know. Guy's name is Booshay or Boshee, something French-sounding. And I think the arms deals activity is ongoing, which might be something for you to pay attention to. You know, leverage." He rubbed his fingers together.

"Bouchet," Katya said. "He made the news a few months ago, but I forget what it was all about. Illegal business dealings of some sort…"

"Like I say, he's into all sorts of stuff. And I'm not the only one doing the digging into ChronoCorp."

"What, have you got some kind of backup spy or something?"

"I've been reading between the lines from my associate, and I'm pretty sure the Imperium has a mole at ChronoCorp."

Everyone in the room froze.

"No…" Xavier said slowly.

"Oh, yeah! Feeding them information on a regular basis."

"Why are you even telling us this?" Katya demanded. "What's in it for you?"

"Pretty simple, really. I've had a lean year, and I'm not so sure based on how they treat me that my buddies at the Primus Imperium are going to make good on all their payments. So I'm willing to do a little work on the side."

"You slime," Todd hissed.

Brad held up his hand. "No, no, let's hear the man out."

"I've been giving it a lot of thought." Rex pressed his fingertips together and walked across the Transport Arena as far as he could—about five paces. "And what I'm thinking is this: I can offer ChronoCorp some great insights into all the little things going on at Primus Imperium. *All* the little things. But I'll have to go undercover."

"Won't someone there recognize you?" Katya asked.

"My dear, I'm a master of disguise. Even my cowboy of an associate—if he's there—won't know who I am. So here's the deal: I'll find out what they're up to and what they plan on doing with those files they had me steal. For your part, ChronoCorp will pay me. Handsomely. There's always an extra charge if I have to go in costume."

"I'm sure there is." Xavier was remarkably calm.

"We can discuss your fee in my office. And I'll have our attorney draw up a quick contract."

"I'm sure you don't need to be so formal."

Xavier's smile was cool. "I'm sure I do."

54

Kelsey and the Burglar

If he was ever going to get ahead financially, Rex Berringer knew he had to take chances, and he was taking one now with ChronoCorp. But he had no doubt that if he supplied Xavier with high-quality surveillance and good intel, he would be paid amply. Xavier was a stand-up guy. Maybe a little unstable with all that twitching and those bloodshot eyes, but reliable nevertheless. Besides, they had a contract.

So it was up to Rex. He had been paying attention to things. He'd noted addresses on his checks and purchase orders, some of which seemed to be shells, but others brick-and-mortar office addresses. He had been to the office once on a drive-by after he had searched online and narrowed down which address was the real center of operations. But it was a small office. He'd also listened as Dresden described various things going on. It all added up to a much larger place, and his money was on a warehouse he knew on the outskirts of town. Could this be the real headquarters of his newly-former-employer?

He used some artfully applied white correction fluid on the ChronoCorp paperwork, transforming it into a new form as he substituted the name and address of the Imperium. In a dim room, it would look perfect. He hoped for a poorly lit reception area.

Rex realized that Kelsey Acres, the name on the inspection forms, could also pass for a woman's name. This would be the key to his success. He had to shave off his only facial hair, that hard-earned moustache, but it would be worth it. Besides, it would only take a couple years to grow it back. As for makeup, both his cosmetics-loving brother and his former wife had coached him on how to look like what his ex called a "Clinique cover girl." That was

M. Carroll, *Plato's Labyrinth*, Science and Fiction, https://doi.org/10.1007/978-3-030-91709-8_54

for a case he had a few years ago. He hoped the rules of fashion hadn't changed too drastically. After his poignant shave and his Maybelline moment, it was time for action.

*   *   *

Ever since his last return from Thera, Xavier had been short on patience and long on adrenaline, and it was getting worse. He couldn't relax, and his focus was gone. He thought about assorted over-the-counter drugs, and some under-the-counter ones. He even considered pot. Marijuana was a dirty habit, and Xavier hadn't partaken since his college days two decades past. But even grass was looking like an option. Colorado had legalized marijuana outlets, but he hadn't gotten a doctor's order for the potent stuff. He would have to get around to that.

He had a contract with Rex Berringer, but he just didn't trust the guy. He didn't need to take anyone else away from their ChronoCorp duties, but he would set out to observe the Imperium/ChronoCorp detective. After he relieved himself.

He looked out the window, across the lot in front of Rex's apartment complex, and peered down the street. How did all those cops on shows do it? They seemed to have bladders of camels as they carried on in their interminable stakeouts. He looked up the street desperately. Gas station? Auto parts store? Nothing. He looked down at his cup holder. There, tilting quietly to one side, was his cup from the morning coffee. The cup was empty—but not for long.

Finally, Rex came out of his apartment. It was well after breakfast, and though Xavier's bladder was now empty, his stomach was gurgling. Would it be worth the wait?

For his part, the PI looked different. He was wearing a wig, walking unsteadily on heels, but Xavier knew the PI's Chevy. It was the only one in the lot with a bumper sticker declaring, "Private Dicks make the best lovers." Xavier guessed that the auto had once been silver or gray, but rust speckled the exterior, the badge of a long, hard life.

Xavier had never covertly followed anyone in a car before. He assumed an important rule was to stay back, to give the mark some space so they didn't spot you. Rex's car belched a plume of blue smoke as he pulled out onto the small side street paralleling his apartment complex. Xavier waited until he was out of direct sight, then pulled out to follow. Rex ran through a yellow light, stranding Xavier at the intersection until the traffic cleared. Xavier moved into the flow and spotted Rex's battered Chevy up ahead. He was driving northwest, toward the industrial part of town.

Xavier kept another car in front of him as Rex made his way into an area of warehouses. Several appeared derelict, and at the end of the block was a taller building that looked just as abandoned. But this one had a sign on it; whatever it was, it was open for business. Rex pulled into a parking space across the street and up half a block, apparently trying his best to be clandestine. Xavier followed suit, parking down the street on the opposite side.

Rex stepped out of his Chevy, adjusted his bra, and brazenly walked up to the dilapidated door. Xavier watched as "the inspector" pulled his familiar clipboard from his jacket. Xavier cracked the window and strained to listen.

"Yeah?" came a gruff voice. "Can I help you?"

"Kelsey Acres, Tri-county health. I assume they called?"

"I'll check. Wait here." Before the door closed completely, the man stuck his head back out and said, "Please."

In what seemed like a lifetime, the man appeared again and let the fidgeting Rex/Kelsey in. The door slammed shut. It would be no problem to sneak in the front door, except for the fact that it had at least two CCTV cameras covering it. There were probably more. The place was far more secure than it looked.

Xavier relaxed his shoulders and crossed the street at an even pace. He was half a block south of the warehouse. He walked up an alley that led to the next street. He could see the warehouse from the corner. There were plenty of cameras on the front side, but less here. And the empty lot between was littered with crates and storage containers. Using the detritus of the area as cover, Xavier ventured to the side of the building. He tried a door, locked. He found another. It was padlocked from the outside. This door had been painted shut, clearly unused and forgotten for some time, but the rusted hinge was barely screwed in. A simple pop and Xavier had the hardware off.

The door opened onto a small storage room. Xavier crept into the gloom. Light shown under the inner door. He opened it quietly, peered out, and saw that the corridor was clear. At the hallway's end, a door opened into some kind of large room. Inside, someone was lecturing.

"But now, in fact we can pluck our law enforcement personnel out of thin air, so to speak."

Someone mumbled a question.

"We don't know yet," said the voice. "But I'm sure the positive changes will far outweigh any collateral damage—as you put it—from quantum effects."

The words filled him with a new level of dread. Another mumbled question drifted down the hall. Xavier was just getting up his courage to sneak closer when a loud klaxon blared. He saw a tiny light blinking above the door, nothing like the ominous ones in movies. He turned to see another like it at the other end of the hall.

Xavier spun around, shut the hall door behind him, and briskly walked back through the storeroom, through the outer door, to freedom. He looked up and stared into a CCTV camera placed just above the entrance. How had me missed that?

The sound continued inside. It was definitely some kind of alarm, and he would lay odds on the cause being Rex Berringer. Xavier got back to his car, started the engine, and let it idle for a moment while he collected his thoughts. So this was it, he thought. Battle lines had been drawn in a war he hadn't known existed. ChronoCorp's secrets were under attack. The first assaults from the fake inspector had been unsuccessful. The threat posed by Rex had been neutralized. He had made his choice, and now he was an ally. An unreliable one. But the Imperium must have others. Worse, somebody in the organization was a mole, but he couldn't figure out who.

The crisis required Xavier's action on several fronts. His initial move would be to convince everyone that they should travel back in time and make changes to put the universe back the way it was. But he doubted that Kat and Todd would agree. In fact, he doubted that it would work. If that course of action was ineffectual, he would remove critical software from the timeline, saving the universe in its current form. First, he would have Emil come up with a virus to wipe the main files at the Imperium. He would do the same for ChronoCorp, although he would move a select few ChronoCorp files from the cloud to portable hard drives, as there might be a place for them in the future as historical artifacts. By the time of the San Francisco meeting, ChronoCorp's secrets would be safer than ever.

The Imperium problem would be dealt with by Emil's virus software, eating its way through the dangerous computer files and leaving nothing left. His adrenaline gave him an edge he hadn't felt in weeks. Finally, he was thinking again.

In the rearview mirror, he spotted movement. He turned around in time to see Rex dashing across the street toward his ancient car. Between the buildings from which Rex had come, a hoard of figures followed. They weren't far behind; Xavier would have to do something. He floored the gas and flipped a U-turn, powering across several yards and ending up even with Rex's car. Through the window, Xavier called, "Get out of here."

The pursuers were upon them. Rex tore away from the curb and motored down the street. Xavier rolled down the far window and called to one of the approaching people. This one looked like a bodyguard from a horror movie.

"Hi guys! Can you tell me where that reptile rescue house is?" Xavier asked cheerily.

Three of the guards promptly skirted his car, looking for Rex.

"You mean Iguana Rescue," the horror movie man said.

"*What?*"

"It's in a private residence just south of town, off 24. There's a lot of glass in the back, like a greenhouse. Look it up. I gotta go."

Xavier watched in bafflement as the man ran after his colleagues. He anticipated that Rex would turn up somewhere safe, with a whole lot of information. The guy was like a cat, always ending up on his feet…and yowling. Mostly, Xavier was surprised by the fact that this world actually had such a thing as an iguana rescue house.

55

Pandora

Rex had one overarching prerequisite for his clientele: that they be able to pay the bills. A moral compass was not required, but if the client had one, it was a nice boon.

Xavier had just extricated him from the Imperium goons. That Bradley paleontologist guy seemed like an okay egg. Katya and the rest of the ChronoCorp gang seemed to lean far more toward the good guy end of the scale than Dresden or the Ambassador.

It felt good working for the guys in the white hats. He had done his share to help them, giving them the heads up on what their competition was up to, and even helping them map out where the Imperium had its facilities. But Rex was no hero. He was a practical businessman. He'd held up his end of the deal with the Imperium, too, providing critical information. That made him a loose end, and people like Dresden tended to tie up loose ends in very dark ways. No matter how noble the cause of ChronoCorp, it was time for Rex Berringer to vamoose. But if he played his cards right, he might be able to get some parting gifts from ChronoCorp first.

* * *

"I almost got caught back there!" Rex hooked a thumb over his shoulder. "You guys should be paying me combat wages."

"It was a little close," Xavier admitted. "But I saved your bacon. Now it's your turn."

Xavier studied Rex as the P.I. wiped lipstick from his mouth and settled into a chair. The conference room was quiet except for the little coffee maker,

© The Author(s), under exclusive license to Springer Nature Switzerland AG 2021
M. Carroll, *Plato's Labyrinth*, Science and Fiction,
https://doi.org/10.1007/978-3-030-91709-8_55

grumbling out something brown and bitter.   "They have a transport pod about like yours, but they also have a thing the size of a San Franciscan cable car. They've got some agenda involving bringing a bunch of soldiers from old times to now, to make some kind of uber-army."

"To what end?" Todd asked.

"I didn't hang around long enough to find out. They discovered the closet door I busted through. I thought it might lead somewhere important. I guess it did, if you collect mops."

"Probably saw you on CCTV, too," Xavier guessed.

"This is the thing we feared most," Todd moaned. He looked at Katya. "Remember when you said that we didn't need to worry unless things got out of hand? Well, they're out of hand."

"Yeah, but what do we do?" Katya asked. "Shut down ChronoCorp and tell the Imperium to do the same out of the goodness of their hearts?"

"Absolutely not!" Xavier thundered. "We still have to present at the conference. It won't hurt to wait another week or so to do something. We can lock down our tech after the conference, but we have lots to talk about first. It's ChronoCorp's moment to shine."

"Even with this alleged army descending on the world?" Katya countered.

Xavier jabbed his hands through empty space. "We can fix this, somehow. Use our own technology. Go back and short-circuit the Imperium's technology. And if that doesn't work, we can do something that involves the life of their leader. Send him on a tangent along the way so he becomes…I don't know…a professional surfer or something."

"You'd just keep going into the past, again and again," Katya frowned, "patching and fixing and ultimately making it worse."

"You're the one who talked about rocking the hourglass like a seesaw," Xavier reminded her. "All that about becoming a great sculptor of time, massaging the past. The great hourglass rocking back and forth, you said. Whatever happened to that?"

"I wasn't thinking of consequences in the right way. And doesn't that place us firmly in the old camp of going back to murder Hitler in 1938? Or of warning the captain of the Titanic to take a more southerly route?" Katya pursed her lips. "The universe would be a hot mess. No, this must be a dead-end, or worse, madness. We've cracked Pandora's lid, and we've got to slam it shut." Katya looked pleadingly at Todd.

Todd closed his eyes for a moment, nodded, and said, "Katya's right. I think we've all proven that it's just too much power to give a handful of researchers, even if we mean well. It goes against human nature. We just can't be responsible with it."

"I just hope we haven't started something we can't stop," Katya said.

Xavier quaked but said nothing. They were ganging up on him. If no one would help him, he would help himself.

Rex finally left for home, and a few minutes later Todd and Katya followed suit, abandoning Xavier to his own unsettled thoughts. What if he could track down Arvin Devereaux at some past public meeting or symposium? He could end it then and there. The worst that could happen was that he would get caught and killed by some gun-happy security guard. Maybe he would be dealt a better hand in the next go-round, if there was one. But he doubted there would be a next time. Some of his friends were Hindus, and they believed you would just go around again, with your karma catching up to you. Todd thought there was one chance, with a benevolent god or a god of retribution or something in between. But for his part, Xavier had lost all faith. He was done. Done with the politics and the hypocrisy and the stupidity of people. He felt desperately empty. A flicker of doubt flashed through his mind: why did the world seem so dark? His outlook had changed, ever since the last Q-slip.

Be that as it may, he knew what he had to do. It would involve some dicey mathematics and variables. Still, it would be worth the risk. He needed to do some talking to himself.

56

Xavier and Xavier

Xavier had found encouragement in Emil's calculations. Perhaps, just perhaps, a Q-slip into the future by a fraction of a second, or a minute, might bring him into contact with a parallel universe—a parallel him. He wasn't insane, was he?

And one approach to this crisis was to talk himself out of something. That would be the simplest. Keep himself from proceeding with the ChronoCorp experiments. If he was successful in that, he could avoid the more extreme approaches he'd been entertaining.

He waited until well after quitting time. Todd had remained until late, working on his presentation. Finally, the building was deserted, and Xavier crept into the Transport Arena. Time for another Q-slip.

Oddly, his pod materialized not in the Transport Arena, but just inside Emil's lab. If he didn't want to be discovered, he would have to move fast.

He steadied himself against the doorjamb for a moment as the room spun. Stepping into the hallway, he stole toward the lounge where he knew he would find himself, the other Xavier. He would think of himself as Xavier *a*. After all, he *was* the alpha, the number one, the big cheese, doing all the work, all the thinking. He was the mover and shaker in the situation, the agent of change. It seemed illogical, but the quantum world seemed to run on illogical precepts.

He reached the door, took in a deep breath, and peered around the edge. Hunched over the table was his past self. Or was it his future self? His sense of the progression of time was fraying around the edges, softening like cotton candy on a summer day. The other Xavier, Xavier Ω, looked different, in the same way that a recording sounds different to the one who speaks it. He

M. Carroll, *Plato's Labyrinth*, Science and Fiction,
https://doi.org/10.1007/978-3-030-91709-8_56

looked worn out. Maybe this Xavier *a* could do something about it, stop the unfolding crisis of mental health and historical consistency. He had to try.

"You really shouldn't do that," he said tentatively.

The Xavier at the table stiffened, but didn't turn around to look. He said, "Do what?"

"Travel. Exploration is for teams, not for solos. Look, you. I came to warn you. It will make you crazy. Don't go. You're playing with quantum fire. *We* are. And our work has opened up the way for others, too. Others who aren't so nice."

He watched for some reaction. When there was none, he added a piece of more specific information. "Watch out for the Primus Imperium. They're the ones you need to worry about."

What was the use? Could he change the mind of his past self without committing some paradoxical faux pas? He gave up. As he headed for the Transport Arena, he called toward the door, "And above all, don't go back." He made his way in the direction of his pod. But then he began to wonder: if his suggestions to his alternate self weren't strong enough already, he needed to reinforce them. He headed for the logbook to see if his whisperings to himself had made any changes. He logged on. He scrolled down. The second trip was still scheduled. But wait: according to the log, it had already taken place. Somehow, his interface with the other timeline wasn't adding up. Perhaps the chronology was skewed because of the closeness of the two timelines.

Xavier *a* rushed back to the lounge. His twin was no longer at the table. He could hear himself moaning on the couch, covered up by the blanket Katya had laid upon him. Sweet Katya. Time had run ahead and he was still catching up.

He thought of the Hawkins drawing, and how it had made an impression, but not deeply enough to last. It was time to make enough of a difference that the changes would stick. He leaned over the figure on the couch.

Lying down on the job! Can you hear me? No reaction. *Hey you. Hey Xavier!*

The Xavier on the couch stirred. If this didn't work, he would have to go with his plan B, and it was a plan that scared him to the core. He had never seriously entertained the thought of killing someone, except that one guy in the Tesla who cut him off on I-25. He tried again.

You idiot. I told you not to go back. What are you thinking? There's too much at stake here. We need to rein it in.

The Xavier on the couch remained asleep, disappointingly inert.

It was time to travel in the opposite direction on the timeline, forward. But only for a few milliseconds. Just enough to meet a parallel Xavier, if there was such a thing. If Emil's alternate math worked. *If, if, if.*

Xavier *a* knew what going backward was like. His encounters with his alternate self were phantasms, dreamlike visitations. But unlike his ghostly character, he hoped this new Xavier—the one moving the opposite direction in time—would be flesh and bone, solid enough to carry out his terrifying plan B.

He was reeling with déjà vu. He could no longer tell what really happened in his past. What had disappeared? What was new? He set off for the lab where he had left his pod, in hopes that his next stop would be home.

* * *

Xavier's permanent migraine was cutting into his thought processes. He was finding it hard to keep his car between the lines or remember which landmark came next. He found himself frequently getting lost. Fortunately, some powerful pain medication awaited him at the local pharmacy, courtesy of his general physician. He grabbed some over-the-counter tranquilizers and stepped to the counter.

"Prescription for Xavier Stengel," he said, flinching at the sound of his own voice. The pharmacist rang up his order. She handed him the little white paper bag with the two medications. Xavier peered into the top.

"I remember these! We used to get these at the drug store when I was a kid; dinosaurs and elephants and flamingos. Haven't seen them in ages. This one's even pink!"

Xavier turned to leave. As he passed through the glass door, he saw the reflection of the baffled pharmacist at the counter behind him. He climbed into his car, considering the little toy hippo at the bottom of the bag.

He pulled into his driveway and entered his darkened home, leaving the lights low. He popped several pills and then fished the pink toy from the bag. He held it up in the dim light. Something caught his attention on the windowsill, a collection of tiny objects. He knew immediately what they were. He added his new pink hippo to the parade of rubber animals that had never been there before.

57

No-Man's-Land

"So, Kat, I've asked for some time off." Brad looked proud of himself, shining out from her tablet. "You've given me the ancient civilization bug. I'm looking into signing up for the conference so I can see your presentation. All the ChronoCorp stuff, that is."

A look of embarrassment shadowed his face for a fleeting moment. "You wouldn't mind, would you?" he asked.

"You *should* come," Katya said. "You'll enjoy it." She had her tablet braced against a saltshaker on her kitchen counter.

A rattling noise sounded from Kat's end, and Brad peered into the camera. "Are you alone?"

"No," she said. "I've got a handsome man at my feet, where all handsome men belong, and he's chowing on tuna kibble. Wellington, say hello."

She leaned over and held the cat up to the screen. "But otherwise, yes."

He smiled weakly. "So before I go racing off to San Francisco with you guys, there are a few loose ends I want to tie up."

"I have no doubt. I'm still tying them up myself."

Brad nodded. "Studies of ancient life tend to be quite linear. This species rises and falls, and then another, and another. And even though the changes aren't always linear, there is a cadence to them. There doesn't seem to be a cadence to this quantum stuff."

"You might be surprised," she teased, scratching the top of Wellington's head. "There's an echo across the universe. Did you know?"

"What sort of echo?" Brad said skeptically.

"It's called three-degree background radiation. The cosmic microwave background. It's an echo of the Big Bang, a leftover fingerprint. We were

© The Author(s), under exclusive license to Springer Nature Switzerland AG 2021
M. Carroll, *Plato's Labyrinth*, Science and Fiction,
https://doi.org/10.1007/978-3-030-91709-8_57

studying it before we tumbled onto the whole temporal travel thing. Seems to me that nature has echoes in other places. Maybe time also has such echoes."

"Echoes in time. Like sound waves bouncing off of things. Interesting thought."

"Speaking of time: be sure you get to the conference on Thursday. Our session is the very first on Friday. Todd goes at 9 am, and I'm up at 10:15. Press conference is at 11:30. They always do that. Wouldn't you think they'd let the speakers eat? How about dinner Thursday night?"

"Dessert. I've got a late flight."

"Dessert. See you at the center's grille at eight? Nine?"

"Eight is good."

They signed off, but the internal conversation was far from over. Katya thought about how she would explain various things to him, and how badly she would like to shift the chatter to more personal things. But this high-profile conference was probably not the place. Sometimes, life simply got too busy.

* * *

Emile's Mobius time loop was quite a different experience from what Xavier was used to. The nearly instant transfer common to the other slips extended into a long (seconds? minutes?) tumble through darkness. But the darkness was not complete: against the void glowed points of multicolored lights streaking across his vision in random directions, each with a comet-like tail. He heard echoes of voices, incoherent mutterings. He caught a phrase in his own voice: "This is Kitane, a woman of some influence." He heard laughter. Or was it the call of a coyote? And another word that was the Minoan for something akin to alarm or danger.

In this twisted version of a quantum leap, Emil had warned him that the talisman might not bring Xavier back to his point of origin. But he had to risk it for the good of his own universe.

The darkness cleared. The lights faded away. Xavier stood within an environment that was at once recognizable landscape and something completely foreign. His mind tried to make sense of it, but he felt as though he was on a ladder with no rungs. The closest he could come was a desolate, open plain, but the ground sloped at an extreme angle. He held out his hand, steadying himself, and realized that it was he who was at an angle, not the world. He stood, swaying. He felt exposed, unprotected in this surreal no-man's-land between universes. Shafts of glistening black stone—obsidian?—pierced the rolling landscape. Their pillars created a rocky labyrinth snaking toward a distant rise, topped by an impossibly tall obelisk. Clouds of white smoke drew a banner from the monolith across a purple sky. Thunder rumbled above, and within came the faint resonance of voices: "Iguanodon…Megalosaurus…Sydenham…Boss Tweed…" Another voice— his own—intoned softly, "It's Todd. He's gone." The word *gone* echoed as it faded into another phrase: "Indicators are set for Thera."

From the corner of his eye, a green, scaly beast galloped out of view. The turpentine aroma of Mediterranean pines, the rising wave of wildflower incense, reminded him of his trips to Thera. Xavier had the overwhelming feeling that his senses betrayed him. Perhaps his perception was not tuned to comprehending this new universe. After all, he had spent a lifetime in a universe that operated by different rules. Why wouldn't this experience seem alien in some ways?

He tried to ignore his surroundings and concentrate. Now, facing himself, he would see whether his Plan B would work. He was beside himself, literally. The other Xavier, which he still thought of as Xavier Ω, stood with his back to him. Ω turned and casually said, "Oh, hello."

He looked like a twin, but at the same time there was something different in his essence. The obelisk had vanished, and the sky was fading into a vaguely familiar checkerboard. Where had he seen it? The glistening rock had lost its deep hues, shifting to shades of gray and chrome. The wildflower fragrance waned, replaced by plastic, ozone, carpet, and burned coffee. As Xavier had suspected, his universe was somehow coalescing, returning to its former state.

"Greetings," Xavier *a* offered.

The man frowned. "Who are you?"

The checkerboard had morphed into ceiling tiles in Xavier's private office. The stony shafts were gone, or rather had metamorphosed into furniture that seemed to be melting back into the forms he knew. Xavier Ω didn't appear to notice the changes. Either this guy was a pancake short of a stack, or there was some sort of disconnect between Xavier *a*'s reality and his.

"Who I am does not matter," Xavier *a* said, "but I have something for you to do. Something very important. I'll provide you with everything you need, including a round-trip plane ticket, and I'll meet you again when you're finished. I'll take you to a little place that is safe. Up the Poudre Canyon. Does that sound all right?"

After a pause, Xavier Ω seemed to focus his attention. He said, "Happy to help out."

* * *

The Mission Bay Conference Center rose from the manicured lawns and low-lying buildings of the campus at the University of California, San Francisco. Mission Bay surrounded the area with its elegant restaurants, high-end retail shops, and luxury condominiums. The Conference Center had its own eatery, simply known as "The Pub." The little grill provided plenty of dessert choices for two hungry travelers.

Kat was already waiting at a table on the patio when Brad stepped in. Her silhouetted profile against the evening lights of San Francisco reminded him of a graceful crane. She had the rich complexion of her father, but she must have gotten the grace from her mother's side. (Brad had seen Ajit in the field; the term "grace" did not come to mind.)

He felt a charge in the air. He had been to many conferences, presented at several, even chaired one or two. But an archaeology-with-physics conference? That was something new. Just what he needed for a change of pace.

Yes, there was electricity, but as Kat smiled at his arrival, he began to wonder if that spark was about more than the upcoming meetings. He thought about his accidental voyage to ancient Thera, and how he wasn't worried about himself. His only thought had been, *what if I never see Katya Joshi again?*

"Hey you!" Brad said.

"Hey yourself. Glad you made it."

Suddenly finding himself only inches away from Katya, he caught a whiff of apricot or peach. The aroma seemed to enhance the ferocious blue of her eyes, the same profound blue he had seen in the waters of the Aegean, the same hue the Colorado skies took on from a high peak, a blue that could pull you in and let you drift. Her eyes reflected her essence: wild, complex, intense. She had been speaking. What was she saying?

"…so I thought tomorrow you could sit at our team's table before our lecture series. I asked. There's room."

"Your team's table? ChronoCorp? Sure. Love to."

"Then the only thing to decide is what kind of treats we're going to munch on tonight."

"Can you order me a hot fudge sundae? I haven't stopped at a restroom since my flight."

"Oh, sure. Down the hall on the left. I know from experience. Would you like wine with that sundae?"

"Something white would be great. Or are you supposed to have red with ice cream?"

He returned as quickly as he could, sat down and spread the napkin on his lap. He realized that his zipper was still down. He usually didn't make that mistake. He reached down surreptitiously and fixed his pants.

"So," Katya said, "Todd will cover the architecture and fashion for the archeologist set, and I will present some of the technical approaches to our quantum travel, something you are all too familiar with."

"You got that right. If there is a next time, I'd rather be going with you than with Rex."

"I'm flattered."

The sundaes and wine came, but the two of them were travel-weary. By the time the sundaes were gone and the wine was soaking in, Brad let out a long yawn and said, "As much as I love the ambience of this fine grille, and the company—of course—I'm bushed."

"Me, too. Thanks for the dessert, and the sparkling conversation—of course."

"Of course," he repeated with a tired smile. "Sleep well."

He stood up. Only too late did he realize that he had zipped the tablecloth into his pants. Silverware clattered to the floor, a glass broke, and what was left of Katya's wine splattered down the front of her blouse.

Katya stood with a quick intake of air. "Well, as they say, dessert's on me."

"I've always wanted to do that."

Katya's eyebrows went up. "Do what?"

"See if I could pull a tablecloth off a table without making anything fall off. Now we know," he blushed.

* * *

In light of the ChronoCorp press conference a few months earlier, Todd Tanaka's presentation was eagerly anticipated. Attendees packed the central room of the Robertson Auditorium to standing room only, and a parade of observers lined the back wall. Todd twisted around in his chair and whispered to Katya, "Now, *that's* intimidating."

"I'm sure they all came early to get good seats for my talk," Katya quipped. "You may have images of ancient Thera, but I've got equations."

He chuckled. The facilitator introduced him, and he took the stage. The screen behind him flared to life, and so did the attendees.

Todd introduced his presentation with some background. "With its palaces, sculpture, murals and shipping, the Minoan empire represents the first advanced civilization in Europe. For our research, we've focused on the meeting of two periods, the MM III Neopalatial and the Late Minoan periods."

Todd continued with some technical data, and then the images began. "The palace we visited was luxurious. Low walls containing spectacular murals formed the boundaries of the paved central and west courtyards. The westernmost court was distinct in that at its northern edge, wide stairs swept downward. We think this was a theater. Sorry about the photo. The light was not great."

Laughter rippled throughout the auditorium.

"Below that court was another level, a repository where we saw clay jars filled with gold and silver, lapis lazuli, carnelian, amethyst, wine and olive oil. Most importantly, several had inscriptions written in Linear A, which we documented. These records will go a long way in breaking the code of the written Minoan language. We also had the privilege of attending a meal. Our hosts served us an assortment of dishes, as you see here."

Todd revealed one of the snapshots that Xavier had taken just before the Minoan chef had absconded with Todd's own bioimager. He did not share that story.

Kat looked at Brad, who was riveted to the screen. "Look familiar?"

He opened his mouth, but no words came out. His eyes were the size of Wellington's food dish.

Todd's lecture seemed to rocket along, and in no time it was her turn. She took to the podium and, as promised, flashed a few equations on the screen. But not many. She displayed images of the pods, the talismans, and some of the other equipment. The ring laser tower was proprietary, but Xavier had given her permission to include the other images. In fact, he had encouraged her to share them—subjects that only weeks ago were considered top secret. She wondered what he was up to.

58

Announcements and Partings

ChronoCorp's eagerly awaited press conference opened with a series of photos, not only of ancient Thera but of nineteenth-century New York City. Once the two dozen images had cycled, Xavier stationed himself behind the podium.

"Ladies and gentlemen, colleagues and friends!" His tone took Katya by surprise. It reminded her not of a researcher presenting results to a distinguished gathering, but rather a cheap carnival huckster selling tickets to see the Bearded Lady. She wondered if he had been taking whatever Todd had been on. "I'm sure you realize by now the power that our new technology offers to the fields of archeology and anthropology, not to mention quantum physics and relativity research. With power comes responsibility. Clap if you agree."

A scattering of nervous applause followed.

"Yes. Yes!" he encouraged. "We at ChronoCorp have delved into the mysteries of the cosmos, the heart of the quantum world, and we have revealed some of its secrets. These revelations have resulted in our capacity to carry out our Q-slips, or time voyages."

Katya shifted in her chair. He was using hyperbole where factual language would be better received.

"We have peered into the very heart of the Minoan civilization, and our work has put ancient Greek studies ahead by decades. Centuries!" He paused, as if waiting for more applause.

Katya leaned over to Todd. "This is a little over the top."

"It's almost as if he wants to be noticed," he said apprehensively.

M. Carroll, *Plato's Labyrinth*, Science and Fiction,
https://doi.org/10.1007/978-3-030-91709-8_58

"But at what cost?" Xavier intoned dramatically. "The power we have is an unknown quantity, and it may be dangerous to the universe itself. We have considered this carefully."

Katya glanced at Todd. His face had drained of color. They knew what was coming.

"It is with regret, therefore, that we announce the complete cessation of our time travel experiments, at least for the foreseeable future." The room filled with muttered amazement. "Foreseeable future," he looked down at the podium, laughing at his unintentional joke. Squinting out at the crowded auditorium, he continued, "We will maintain our research into the nature and ramifications of the work, but for now, we consider this the only prudent course of action. We would also encourage any scientists pursuing this avenue of research to stand down."

The moderator stepped onto the stage and said something inaudible to Xavier. Xavier came down and sat beside Todd. He grinned wildly and whispered, "What did you think?"

Katya opened her mouth to respond, but the woman at the front podium began to speak.

"In lieu of a Q and A session, I have an important announcement to make. Late this morning, our esteemed colleague Arvin Devereaux was nearly assassinated at his home in Michigan. At this time, the assailant is unknown, but the authorities are treating the incident as attempted murder. Dr. Devereaux was shot, but not fatally." A shocked hush had fallen over the auditorium. "He is recovering in the hospital in a secured room."

The moderator dismissed the morning sessions, and people began to mill as they headed for lunch. The ChronoCorp group proceeded to the lobby.

"Imagine that," Xavier murmured.

"What, that he survived?" Todd snapped.

Xavier tittered strangely. "No, of course not." Katya could smell his stale breath, see the sheen of sweat on his brow. His lips peeled back in a grimace.

"At any rate, it sounds like their CCTV wasn't enough to show the culprit," Todd suggested. "Maybe they didn't have any CCTV."

"Oh, they do," Xavier said. A look of panic flickered across his face, and he continued, "I mean, wouldn't you? Wealthy man? Big house."

"Makes sense," Kat said.

Xavier glanced at his phone. "Todd, we gotta go catch our flight. Don't forget, everyone, quick meeting to debrief at five-thirty tonight. I know it's a push and we'll all be tired, but don't be late."

"I'll be hot on your heels," Katya said.

Xavier sauntered across the lobby to the concierge. Katya held back and turned to Todd.

"I don't think the man's well, Todd. Too many trips. And I'm not talking about the airlines."

"The thought had crossed my mind. I'm afraid he's losing his perspective. He's becoming unhinged or something. But let's regroup later, sit down with him and finally talk it out. I gotta go," He said, waving goodbye as he rushed across the lobby.

Brad waited until Xavier and Todd had left to approach Katya. "Everything all right?"

"Yeah," she said, gazing in the direction of their retreat. "Yes, it is."

"Should we grab a bite?" Brad asked. "We've got time until our flights, and I heard about a great place up on the Embarcadero."

They took a cab to a nice little bistro with a view of the Oakland Bay Bridge to the south, and Yerba Buena and Treasure Islands to the northeast. Through the windows on the far side of the restaurant, the famous Coit Tower loomed out of the mist.

Lunch came. Brad spoke around a spoonful of cioppino. "Want some?"

Kat screwed up her face. "I always thought they put a few too many things in there. It smells like a locker room. And you have to avoid all those shells."

"Wait—you're not supposed to eat the shells? No wonder it gives me heartburn." He held up his spoon in salute. "I really enjoyed this meeting. These few days have had some amazing revelations."

"Not just for you," she said. "I felt a little blindsided by Xavier's announcement. He's been acting strangely this whole trip."

"Yeah, I got that feeling."

"But Xavier had some good reasons for putting things on hold."

Brad soaked up cioppino juice into a crust of bread. "You mean, like a madman trying to take over the world with similar tech?"

Kat took in a deep breath. "I think the head of the Primus Imperium, whoever he is, is anything but mad. His actions strike me as those of a calculating, organized mind."

"You can be calculating and organized and still be mad as a hatter."

"Maybe so. But I think I—we—need to take some proactive measures."

"Sounds dangerous." Brad didn't try to hide his genuine concern.

"It could be. But we'll be careful. We're a good team."

"Maybe I can help?"

"Best leave it to us experts."

"Like I said, a day of revelations." He knew it was time to let it go, so he changed the subject. But he would return to it later. He didn't want Katya running off half-cocked on a mission to save the world. With all the nonchalance he could muster, he said, "So what's on your agenda when you get back?"

"Meeting tonight to debrief. Then, oddly enough, Mila van Dijk wants to see me about something. When I asked what it was all about, she was very hush hush. It was a bit unlike her. Guess I'll just have to wait and see. Speaking of revelations…" Katya glanced to the side, and then leaned toward Brad. "I might as well tell you that Mila has it bad for Xavier."

"Has it bad? Not sure I follow."

"As in a high school crush. She adores him. She's been asking all sorts of things about his relationships and his work habits and hobbies and the like."

"I never would have guessed. Philanthropist/archeology buff meets quantum physicist. Who knew?" He wiped his mouth with his napkin, which was thoroughly soaked in orange fish and clam juices.

"What's on your horizon?" Katya asked.

"Back to the museum to make up for lost time."

Katya looked at her phone. "Jeez, it got late! My flight's going to leave without me."

"Can they even do that to important people like you?"

She threw her napkin at him.

*   *   *

Katya left for the airport. Brad's flight didn't leave for another three hours, and he didn't like waiting in airports, so he ducked into the hotel bar and ordered a Shirley Temple.

"Seems a little early in the day for such strong stuff," a man on the next stool over said amicably.

"They mix these with an extra shot of Ginger Ale here. That's really why I flew to this conference." The two introduced themselves, and Brad added, "Didn't you present a paper yesterday on house murals?"

The man nodded.

"You had some beautiful shots."

The archeologist glared at his drink. "I don't think you've been paying attention. Didn't you see the ChronoCorp images? *Contemporary* photos? If this keeps up, traditional archeology may be out of business."

"Are you saying archeology will be a thing of the past?" Brad said with a grin.

"Funny," the man said without humor.

"I think you'll be all right," Brad said. "I have a friend who's an artist. When digital art really got going, people said that traditional art was on its way out. But it wasn't. It just changed to adapt. I'm sure you'll do the same."

The man grunted and asked the bartender for another. "You know what really irks me? How much funding it took for them to do all that. I scrape by, year after year, with grants so small that I can only support one and a half grad students at most. But they find Mila van Dijk and plug into her zillions, all because she loves the ancient Aegean and she got all that money from her husband."

"Her husband?"

"Ex, actually." The man took a sip of his newly filled drink. "Yeah, got most of her money from her ex-husband's arms deals, I bet."

"Arms deals?" Brad felt as if someone had poured his Shirley Temple down the back of his neck.

"Oh, yeah. Franklin Louis Bouchet. He's a real piece of work. The guy made his mark selling RPGs and automatic weapons on the sly, but the government caught up with him. Threw him in the pokey for a while. Guess he got reformed."

Could this be the same Bouchet Rex had mentioned? Brad wondered.

"Of course," the man added, "van Dijk went by a different name back then. Remade her whole coiffure when she arrived on the jet set scene."

If the Ambassador was Bouchet, and if Bouchet had been married to Mila van Dijk, was Katya walking into a trap? Brad had to get in touch. As he sprang to his feet, the other man said, "Have a nice flight."

"You, too," Brad said.

He punched in Katya's number, but it went directly to voicemail. "Kat, check your texts right away."

He texted her.

THE AMBASSADOR AND MILA MAY BE INVOLVED TOGETHER IN SOMETHING. WATCH YOUR BACK, KAT. I REALLY CARE FOR YOU, AND I DON'T WANT ANYTHING TO—

He deleted the last phrase, leaving it at "Watch your back, Kat." He hit send, checked the time, and called his airline to change his flight destination. Brad was going back to Colorado.

59

Missing in Action

Kat shut off her cell service, shuttered her tablet, and flew in glorious silence until she reached her destination. Her phone was still on airplane mode when she arrived at ChronoCorp. Xavier met her at the door.

"You're here," she said cheerfully.

"Where have you been?" he snarled.

"Forty thousand feet above the western US, like you."

"Of course. Sorry. I tried your phone, but—"

"Oops. Are we still meeting?" she asked as she turned her service back on. A long line of messages from Xavier paraded down the screen, followed by another string from Brad.

"It's Todd," Xavier said, pacing around in a small circle like a caged Whirling Dervish. "He's gone."

"Gone? What is this, freelance week? Everybody's running off with impromptu trips. Brad, Rex, now Todd."

"I think he's having some kind of breakdown."

You're one to talk, she thought. The man was sweating like a stuck pig, his face flushed and his movements jerky. "Where could he have gone?" Xavier murmured.

Katya thought back to Todd's desktop image of *Le Parisienne*. "Oh."

"What?" Xavier thundered.

"Did you notice, at the conference, when you made your announcement about our hiatus? I see it now: I think he's gone back, to see Kitane again."

Xavier dashed down the corridor toward the Transport Arena, with Katya close behind. As they entered, he went over to a monitor and began tapping keys. Katya examined the pods.

© The Author(s), under exclusive license to Springer Nature Switzerland AG 2021
M. Carroll, *Plato's Labyrinth*, Science and Fiction,
https://doi.org/10.1007/978-3-030-91709-8_59

"The indicators are set for Thera, all right," Xavier said. "Talk about a long-distance relationship."

Katya started as she looked at the wall. Hanging in their cradles were all four control units. "But he left without his talisman!"

"A one-way trip."

"I'll bet he just wasn't thinking straight." Katya said.

"If you're going back for him, make it count. Bring one of Emil's new bio-imagers. It records sound. He'll show you."

"He's still here too?"

"Dedicated employee." Xavier hit his earpiece and summoned the tech.

Emil arrived in moments, rushing into the small Arena in his stereotypical lab coat. He gave Katya a quick overview of his improvements. Emil stroked the imager as if it were a pet hedgehog. "Now, remember, it's either/or. You can take still images or you can record, but not both at once."

"Emil, will you see to Miss Joshi's departure and return?"

"Sure, Boss."

"Kat," Xavier said, "I won't be here when you get back. I have a trip to go on. Maybe a couple. But when this is all over, things will be better. Everything will. Don't be long, and be careful."

Katya was too distracted to think much about his words as she grabbed two talismans from the wall. Emil helped her into the pod. She braced herself for another voyage.

*   *   *

After all that had transpired with Xavier and Todd and the fallout from their Q-slips, Katya knew that going back again might well send her on an inexorable course toward insanity; at the very least, it wasn't very good for her long-term health. And now, as she sat on the edge of the pod, her body confirmed it. Every sinew, every nerve, every fiber right down to her gut told her it was so. Deep within her core, a greasy, malignant wave surged up. Her vision blurred, but she had a job to do. She took a breath, then another, willing herself to calm the whirlwind in her chest.

Katya's mind cleared, but she didn't move for a few moments. She listened to the gulls and took in the Aegean air. The air was rank with an alarming mix of sulfur and smoke.

She wondered where Todd could have gone. Time's currents may have brought her to a site near his arrival point, but he wouldn't have stayed there. He was on a mission. That mission involved Kitane, so she would have to find the priestess if she was to find Todd. How much time had elapsed in Thera? Its passage here might not directly parallel time at ChronoCorp. Would a week have gone by since Todd arrived? A month?

Outside the pod, she heard what sounded like thunder. She opened the canopy and sat up, then stood. The sky toward Thera's central mountain had darkened into an angry maelstrom. She spun around toward the coastline below. There, sitting on a rock a few yards down, Todd slouched over, clutching something to his chest. He looked old.

Katya walked down to him, gingerly putting a hand on his shoulder. "Todd! Are you okay?"

Todd hugged himself, his teeth chattering. His eyes had the same glossy pink shade that Xavier's had gotten. "I had to come back. One last time. I had to see her."

Katya hurt for him. "Did you? Find her?"

He canted his head. "She left. On a boat. She gave me this."

He held out a small votive, a figure of a bull leaper that fit in the palm of his hand.

"Beautiful," Katya said.

"She asked me to come, but I knew I couldn't. We're not welcome here, in this time, in this place. There's no sanctuary for us. Maybe it's in our genes. We're just not designed to be here. Do you feel it?"

"Yes, I do. I did from my first trip. It feels wrong." Katya quieted and let Todd be with his thoughts for a moment. She said, "I'm sorry you had to leave her."

Todd raised his eyes to the striking ultramarine still lining the horizon. His eyes were rheumy. His shoulders quaked. He sniffed. "Me, too." Todd rocked back and forth. He seemed to study that impossibly blue edge of the world. "It's surrounding us, you know. Closing in like a net: The Eternal. It's so big, all that was before the beginning of all things. And it will be here after us. We think we're so big, so important. We've broken into the fourth dimension. But we had no idea, did we? No idea how big it was or how powerful or moving. It's all probably above the paygrade of the human race." His breathing became labored, quickened. "And us? We are nothing. A scattering of flotsam. A faint candle flicker before an exploding sun."

Todd's doomsday speech alarmed her, but she tried not to show it. She patted him firmly on the knee. "How about a little distraction? A quick mission before we go back."

"Mission?" He closed his eyes and smiled. "Back to that temple?"

"I just can't resist. There's a room at the back that's been giving me nightmares."

"I'd like to see it. Ironic that I've been here four times now and never seen it. Too busy with the palaces."

"And you got great data on them. You're a good researcher. It's this way."

Todd stood as if he had downed half a bottle of Jack Daniels. Katya steadied him. They climbed the crest of the hill and turned east, toward the site where they had first arrived, back at the beginning of this great adventure.

The steep climb winded Todd, but the rumblings of Thera and the lowering sky urged them on. The town square was silent. Windows stared out at them, dark and empty eyes hushed by the great exodus of the Minoans. Even the fountain at the center of the plaza seemed to whisper.

They stepped through the temple entrance. No lamps lit the interior, but gray light cascaded through the roof's light wells. Todd walked to the little altar and dragged a finger along its lip. Katya remembered Mila van Dijk making a similar gesture along the Minoan treasure cases of her mansion. Todd studied the murals.

"Hey, these look just like Mila's mural sections."

"*Just* like," Katya said.

Todd came up to her, carrying something folded. "I found these in the corner. You'll look like a proper Minoan priestess."

"Thanks, Todd. Now, send your eyeballs somewhere else while I transform." Katya dressed quickly—she remembered how to wrap the clothes around her from her other visits—and dashed to the back of the temple. She pulled the curtain aside, then stood at the door, baffled. Beyond it lay a solid stone wall, slightly curved outward. She dropped the drape back into place.

"They couldn't have come through solid brick," she muttered. She spotted the embossed face of a stone, protruding subtly from the surface of the wall. She pulled on one edge. Nothing happened. She pushed against it. A low growl rumbled from behind the masonry, a deep grinding. The curtain wavered gently, and light flowed through the thin opening at its edge. Katya pulled the fabric to the side. The stonework was gone, replaced by a doorway opening on to the small room she had seen before. On the back wall, the mural she remembered was gone, replaced by a painting of sea lilies.

"Do you hear that?" Todd asked, coming up beside her. "That hissing?"

"There, at the floor." She pointed to a small pipe extending from the base of the outer wall. "Could it be some kind of steam heating?"

"This room seems plenty warm. Maybe they're using steam for some kind of power. To open this doorway."

Todd and Katya stepped through the entry. Katya spotted another embossed stone. She pressed. Suddenly, the entire room began to move. They both spread their feet and stuck their arms out defensively, braced for the movement. She thought back to the ChronoCorp elevator, but this room did not ascend. Instead, it rotated. They could hear stone against stone outside the curtain and walls. The inner wall rotated more slowly than the floor they stood on, and in moments an entryway appeared in it. The chamber slid to a stop. The aperture was dark.

Katya looked at Todd.

"Whiskey Tango Foxtrot," Todd whispered.

"A steam-powered room. We're a long way from Fulton." Katya stepped over to the darkened access and peered in.

"I see light down there. And steps."

"Should we go down?"

"It's our last chance to find out what this place is."

"Storage, I'm sure," Todd said flatly. "All the palaces and most temples they've excavated have had storage below. Lots of pithoi and so forth."

"But why would they have such a complex entry system? We have to check."

"Sure," he said with a quaver in his voice.

"You gonna be okay?" she asked.

He gave her a thumbs-up.

She instinctively reached for her phone's flashlight, but of course it wasn't there; she had left it at the lab. A time-travel-dissolved phone was the last thing she needed. She waited until her eyes adjusted to the dimly lit staircase, then ventured below. There were only five steps, with a wall along one side to lean against. As she reached the floor, she realized that the light was being shunted through a series of light wells far above. Clever.

Todd descended the steps precariously. Once down, he said, "They sure weren't big on indoor lighting. I'll bet they had cheap energy bills, though."

"And free hot water," Katya added.

One lamp burned on a distant wall. They followed its light into the darkness of the room's far end. As they approached, Katya could hear a sort of ticking sound. A lower tone, nearly inaudible, played an underscore and tingled within their chests.

The first thing Katya saw was a heavy rope leading to the ceiling. The rope fed through a rounded hook, and on its end was tied a large stone weight. A second weight hung behind. Katya followed the ropes with her eyes. They dropped down to what she thought was a poorly made table. She struggled to see the details, and when she did she was even more confused. The table was actually a series of flat, interlocked bronze gears and armatures. The dining-room-table-sized construction looked familiar. She had seen one just like it somewhere. Finally, she looked at the mystifying object with a new perspective. This was a horizontal version of what she had seen before. If she was right, this would reset the ancient history books and rock the foundations of archeology itself.

"Todd, it's…it's…the Antikythera mechanism."

"Can't be. I saw that thing in the Athens museum. First analog computer? Seventy-some interlocking gears? But it was the size of a shoebox. And this thing is fifteen hundred years older. I suppose we lost a lot of history in the Theran explosion, but…it can't be, can it?"

Could it be? She gazed. It was majestic, with its ticking machineries, its whirring unnamed apparatuses, its stairstep contrivances. The great gears turned in the darkness, marking years and seasons, eclipses and harvests, inexorable time. It was all about time here, in this ancient chamber with its ancient mechanism, as much as the Transport Arena was about time. Here was more than a calculator for farming, more than a gizmo to chart the seasons. The assemblage embodied greater mystery, as if its inner workings held the essence of the clockwork cosmos in all its spinning worlds, pinwheeling galaxies, and exploding stars. It played the music of the spheres.

A voice came from behind them. "Antikythera?"

They wheeled around to see an ancient woman, her sun-darkened face deeply creased and her hair snowy white. She appeared to have no teeth, but her eyes shimmered with wisdom and mirth, even now. She wore the garb of a priestess, a similar outfit to what Kitane had worn in her temple. She lit an oil lamp and held it up to look at them. Katya could see the glistening of metal wheels within wheels on the device beyond.

The woman exuded a deep calmness. She had not been left behind, Katya sensed. Rather, she had remained behind of her own will. Perhaps the sea voyage would prove too much for her. Maybe she had lived her entire life here and wanted to die here, too.

The woman began speaking. All they caught was the word Antikythera again. Katya slapped her forehead. She pulled the bioimager out and began recording the audio.

The woman continued animatedly for some time, pointing at various gears, at ropes, delicate balances rocking back and forth, interlocking rings engraved with lines and letters and words, at assemblages and the weights swaying overhead.

"Wait a minute; I recognize a word she just said. Kitane used that word for 'again' or 'repeat', something like that."

"Are you catching anything else?"

"A little. There are key words I don't understand."

"Xavier will." She held up the gooey bioimager.

"Hope so. I'm a little bamboozled. I think I still took a few too many meds before this slip."

An earthquake shifted the floor back and forth. A rumble followed, casting some ceiling plaster onto the floor. The old woman waved toward them and then toward the doorway. Clearly she wanted them to flee to safety. They did, but not until Katya had a full set of images, top and bottom, of the strange clockwork computer of another time. While the woman was looking at Todd, Katya snapped a shot of her timeless face, a face of bravery. Why did she stay, while everyone else was gone? Was she the keeper of the device? A priestess who had given an oath to "go down with the ship?"

They climbed the stairs quickly. Katya reached over to the curtain and opened it. Behind it was the solid stone wall.

60

Death-Throes of Thera

The Ambassador addressed his adjutant as if the woman was a schoolgirl.

"Do you know, Miss Graves, why the Roman Empire met with such success? Why they had so many loyal followers, even amongst the peoples they conquered?"

The adjutant did know. In fact, Ancient and Medieval History had been her minor. But she majored in Diplomacy, and she knew that this moment was no time to show off her education. "Why is that?" she asked on cue.

"Once they had conquered someone, they allowed those people to have some degree of self-rule. And they protected them. Helped them set up infrastructure, even. In fact, they endeared the conquered to themselves."

"Ahh, I see," Graves said, trying to see.

"You see, we cannot have ChronoCorp fighting us. They need to be on our side."

A knock at the door announced the arrival of Dresden.

"Come in, come in," the Ambassador said with cool measure. "What do you have for me?"

Dresden consulted a tiny notepad. Flipping through the top few pages, he began to speak.

"Xavier Stengel. Single. No family nearby. Unattached." Dresden said the last with gravity. "He's a self-made man, influential in his field. He won't cave easily. I do have more details."

Dresden closed his little tablet and stuffed it into the breast pocket of his denim jacket.

© The Author(s), under exclusive license to Springer Nature Switzerland AG 2021
M. Carroll, *Plato's Labyrinth*, Science and Fiction,
https://doi.org/10.1007/978-3-030-91709-8_60

"Good. Type them up and email them to me asap. And so you see," the Ambassador said to Graves and Dresden, "we're forced to carry out an intervention. On site."

"An infiltration seems quite dangerous at this juncture," the adjutant suggested. Dresden smiled at her, apparently wondering how long she would last here. Bouchet wondered if the Texan knew she was the Ambassador's niece.

"ChronoCorp has forced our hand. As they made clear at the San Francisco conference, they are deciding how—not if, but how—to stop all of their temporal experiments. If they go public with the 'dangers of time travel' or some such rot, we'll be getting bad PR before our plans can mature. With the intrusion of that strange woman and the strides they've made, and their threat to discontinue, we're up against the wall. It is time for some rather specific action." He turned to the quiet Texan. "Mr. Dresden. I need you to craft a plan to take Todd Tanaka into our safekeeping. As my guest, of course. He will be kept comfortable and secure. And quiet. We'll also need to add Katya Joshi to the guest list."

"Yessir. To what end, may I ask?"

"Bargaining chips. They will soften Stengel's stubborn quality."

"What's our timeline?"

"As soon as you identify a safe opportunity to capture Tanaka, take it."

Dresden nodded, tipped his hat at the adjutant, and left.

"What about Joshi?" the adjutant asked.

"That will be up to our mole."

"You want me to get in touch with her?"

"I'll do it. Circumstances have trapped us, and it's now something she must do. She won't be happy, but that seems to be a permanent state for my dear ex-wife."

"Poor Mila."

"Oh, yes," the Ambassador fluttered his hand dismissively, "Poor, poor Mila."

* * *

Todd rapped his knuckles against the stone wall blocking the doorway. "Solid."

"Let's find out how this thing really works. Hold the curtain back so we can watch."

Todd held the cloth aside. Katya depressed the stone in the wall. The low grinding began again. The stone wall covering the exit slowly slid to the side as the room turned. Masonry slipped ponderously by until another opening appeared. The room or rooms beyond were dark as night.

"That's not what I expected," Todd said. "How do we get out of here?"

Katya held her hand up. "Patience, grasshopper." She pressed the stone and the room began to rotate again. The murals on the back wall changed as the floor, outer wall and ceiling moved. Another door came into view. Through it they could see the temple.

"Just as I thought," Katya said. "Round room. Well, more like a donut, with the middle of the donut locked in place. The rest of the room rotates to expose several different rooms behind. If you want to get out, you just keep pushing that stone until you get to the right exit."

"Why not just build a simple hallway?" Todd asked.

"You would, unless you wanted to keep something very secret." She looked back toward the way they had come, back toward the great clockwork mechanism.

A pipe embedded in the floor exhaled a gust of hot air at Todd. "Definitely steam powered. I wonder how much of this civilization was…"

Another rumble shook the floor. Powder drifted from the ceiling.

"Let's go!" Katya hollered.

Katya stepped through the doorway, into the temple. She looked back through the exit they had just come through. On the wall at the back of the little room, painted dolphins played in the waves. The wall was back where it had been at the beginning.

Through the main entrance they could hear a fusillade of stones falling from the heavens, a spring shower of steaming basalt. Katya's outfit didn't come with shoes, and Todd's had sandals that didn't fit him, so he had thrown them away. Hot pumice scattered across the ground. It would be tough going. The acrid aroma of burning plants and wood hit them. Smoke burned their eyes, and they could feel waves of heat washing up the hill from rivers of incandescent rock. Animals shrieked from the forested hillsides above.

Making their way downhill as quickly as their bare feet would carry them, Todd and Katya reached the pod site. Todd's wheezing alarmed her, and his pallor was corpse-like.

"Todd, let's get you home. Go for it," she encouraged, handing him his talisman.

He triggered the device. After a terrifying pause, a flash of light announced the pod's arrival. Katya helped him into the pod. He remained sitting up for a moment.

"Forget how to work this thing?" she asked.

Todd shivered. Katya put her hand against his forehead. "You're a bit feverish. As soon as you're back, we need to get some Tylenol down you."

"Don't forget," he said through clenched teeth. "I've already got a chemical smorgasbord inside. I'll be fine."

Leaning unsteadily against the pod, he held the little votive to his chest. Katya didn't try to take it away. It would be gone soon enough. He lay down as if it was the last thing he wanted. It probably was. Katya sealed him into his chamber. It began to glow.

A deeper, sonorous crack issued from the summit of Thera's central caldera. Turning back to the spot where Todd's pod had been, Katya felt a rain of hot gravel. Soon it would be raining molten rock. She hit her talisman's trigger. She looked back down the mountain, toward the ships bobbing in the rough waters. Most were headed out to sea, south to Crete or east to Rhodes. Light came from behind, flooding the bushes and rocks before her, announcing the arrival of another pod. She spun around, vaulted for the coffin like an Olympic diver, lay down, and closed the lid. Through the glass she could see the people bolting toward the bay. Most were gone, and they'd taken their treasures with them. They knew the signs. Unlike the unlucky inhabitants of Pompeii, these volcano dwellers had time.

But not all of them. Searing rocks and ash began to fall on the Minoans headed for the beach. A woman sheltered her child with a pillow. A man carried his elderly mother on his back, wavering with each step. The desperate figures reminded Katya of the fairy tale piglets who had built their houses out of straw. Katya looked back toward the summit. The blue of the sky swirled away, eaten by the black cloud expanding higher and higher across the firmament. Within, crackling forks of purple lightning tore through the dark fog, illuminating flying boulders, some of them smoldering red-orange. At any moment, the very fabric of the mountain would disintegrate into incandescent landslides, pouring into the sea, shoving tidal waves throughout the Greek islands and into Asia Minor. She didn't want to be there.

Above her, dropping straight down from the gangrenous sky, a blazing rock the size of a school bus dragged an incandescent tail in its wake. It came closer. Closer. It mesmerized her into paralysis. If she took too long, it would be her last view of Santorini or anywhere else. The boulder was upon her. She had only a moment. She held up her talisman and fumbled for the trigger, but her hand was slick with sweat. As the burning meteor closed the last few yards, the talisman slipped through her fingers.

Katya grabbed blindly for the little control. It was wedged between her and the side of the pod. She yanked on the lanyard and clawed for the trigger. Just above the pod, the mountain of glowing debris blotted out the world around her. She squinted her eyes closed and heard a crash.

61

DOA

Dresden had worked with his share of bizarre bosses before, but the Ambassador was one of the most disturbing. What alarmed Dresden most was the speed with which the Ambassador changed his outlook and, consequently, his strategies. There seemed to be no design here but rather a swift spiral into the paradigm of a maniac. His vision had leapt from corporate espionage to using people as bargaining chips, from kidnapping to murder, from exploration to widespread military action.

Dresden had removed several people from this mortal coil throughout his career; it came with the business he was in. He tried not to think too hard about their personal lives, or the orders coming from above. It had been a long time since he had to resort to anything so crude. Still, he was beyond second thoughts when it came to assassination, as long as it was for a good cause. But the Ambassador wielded impressive—and terrifying—power, and that power was increasing as the technology of the Imperium progressed. Coupled with the Ambassador's recent unstable demeanor, the Primus Imperium represented a dangerous combination, and Dresden had come to believe that this mix was bad for the world at large. It was time for him to get out.

* * *

The phone rang in Xavier's office. It was very late, and he wanted to go home, but he had to mind the store as long as Kat and Todd were out. He considered calling in the full team, but he had to get that aggravating phone first. It was the receptionist. He'd forgotten to tell the poor woman to go home.

"Sir? There's an Ajit Juicy here to see you."

© The Author(s), under exclusive license to Springer Nature Switzerland AG 2021
M. Carroll, *Plato's Labyrinth*, Science and Fiction,
https://doi.org/10.1007/978-3-030 91709-8_61

Xavier could hear Ajit in the background correcting her.

"Josie," she tried again.

"I'll come out," he said, rubbing his eyes. He was in no mood for making excuses about a missing daughter. He didn't have the energy. He was spending all his reserves just dealing with the aftermath of the conference and the botched attempt on the life of Devereaux.

He called Emil, who had gone home on time for a change. Then he called Brianne. This was an emergency. He needed a few techs here in case something went wrong. Time to call in the cavalry, such as they were.

He shuffled to the lobby, told the receptionist they were done for the day, and invited Ajit back to his office where he could monitor the Arena from his workstation.

"Coffee?" he asked. In the corner of his office, a heat-burnished glass pot gurgled.

"No thanks," Ajit said.

"Your daughter isn't here."

"I actually came to talk to you."

Xavier threw him a wary glance. "I see. What about?"

"I heard about ChronoCorp's big announcement. Research on hold. Taking a step back in experimentation."

"Correct."

Ajit scanned his surroundings. "From the look of things here, you're still pretty busy."

"We are. With those things you just mentioned." Xavier pressed. "What did you need to talk to me about?"

"It's about H.G. Wells. Or Michael Crichton or Kim Stanley Robinson, if you prefer."

"I'm afraid I haven't had much time lately to read science fiction."

"Perhaps you should. When Crichton's *Jurassic Park* came out, it was a subject close to my heart."

"I can imagine," Xavier said. *The man's daughter is gone and he's talking world literature.* But he wasn't about to illuminate Ajit on his daughter's possible MIA status.

Ajit brought a finger to his bottom lip. "Yes, it resonated with me, partly because it had so much in common with the writings of Wells. Do you know why?"

Xavier frowned in puzzlement. "Hm, Velociraptors and Martians? No clue."

"You see, those stories weren't really about dinosaur parks and alien invasions. They were about hubris. About human arrogance and a lack of responsibility. Modern-day Towers of Babel."

Xavier shifted uneasily. "I see where you're going with this."

"That's exactly why I came to you. I know you're a good researcher, a responsible one. I would never lecture a fellow researcher, but it's an area that interests me, and I'm sure you've had to struggle with it as well."

"I have had my struggles, personal as well as professional."

"And what of all this?" Ajit swept his hand around the room. "You announced a cessation to the work, but here it is."

"So much progress is represented in this facility." Xavier said it with fondness, as if surveying a room filled with old friends. And he wasn't about to be lectured about how he should be managing the affairs of ChronoCorp. He added, "But we will be discontinuing our Q-slips. Soon."

Ajit looked around, taking in the cubicles, monitors, cables, wall-mounted readouts and other marvels. "Does it give you pause?"

"The coffee maker certainly does, but I'm buying us a new one."

Ajit didn't smile. "So, Xavier, is there an end game to this?"

An end game. Even now, Xavier's plans were progressing. The first stage, the assassination of Arvin Devereaux, had failed on two counts. First, the assassin had not attacked at a point on the timeline far enough in the past, and second, he had missed. In fact, those events in Michigan caused him to wonder about Xavier Ω. The man was so malleable, as if he was just a reflection in the mirror. He didn't seem to comprehend the full picture, seemed to have no interest in the ramifications of his own actions. Maybe it was because of the nearness of the two universes and the fact that the second Xavier was somewhere in between. In fact, Xavier thought that sometimes he glimpsed a shimmering around the edge of his counterpart, as if he wasn't quite all here.

Just two more assassinations: the funding source for ChronoCorp and then the head of the Imperium. Yes, he had to add the guy in charge of Primus Imperium to really get rid of the time travel remnant of the continuum. Then, finally, all would be right as rain.

But what about Emil? He had done all the complicated math. And surely the Ambassador had his own version of Emil, some bright tech who could do it all again and screw up the universe—this universe—a second time. Something would have to be done about that, too. Two more killings? This was getting a little busy. The whole affair was like an onion, but could he ever reach the center? Xavier might have to actually get involved himself, rather than leaving it all to the somewhat unpredictable Xavier Ω. He marveled at the fact that he could even consider murdering someone. When had his outlook on life changed so much?

The intercom beeped and a voice on the other end said, "Energy curve is on the upswing. We've got returning pods."

"I believe your daughter is back."

* * *

The crash Katya heard wasn't in Thera. It was the sound of Todd falling out of his pod in the Transport Arena. As she climbed from her pod, she saw Xavier and Emil bending over him. Emil was doing chest compressions. She spotted her father across the room. He was on the phone, calling for the EMTs.

Though it was only minutes, it seemed to Katya like an eternity. The paramedics took Todd away amid a tangle of tubing and plastic bags.

Ajit stepped in front of Katya and pointed to the door. "Dinner. Now."

Katya Joshi was a grown and independent woman, just as her parents had raised her to be. But no matter the circumstances, Ajit would always be her father. She would always be his daughter. They didn't mind at all. And when he got that tone, she didn't argue.

They returned to Goldsmythe's Pub & Grille. Katya braced herself for a lecture, or at least a barrage of questions. Perhaps she could head her dad off at the pass.

"Todd left without any plan," she said, "Bad things happen when people act on impulse."

"Yes, yes they do. I hope he'll be okay."

"Me, too." She gazed into her coffee cup and studied the swirls of cream. They looked like the colorful storms of Jupiter, graceful and menacing.

"Remember your stagecraft class in high school?" Ajit said.

"Mr. Brookes' class. One of my favorites. We could build an entire cityscape out of wood, or a landscape out of scraps of board and dry wall, but you put a scrim in front of the stage set, and it all evened out and looked soft and coherent and real. Magical. What brought that up?"

"Maybe time is like that, a thin gauze between us in the present and the world of the past." Ajit quieted for a moment, then nodded as if he had made up his mind about something. "I remember an Impressionist exhibit your mother and I went to. We stepped through the door to this room and there, just around the corner, was a large Monet oil. The colors were vibrant, but those ropy brush strokes didn't make much sense. A large sweep of purples and greens here, some brighter lavender and blue there, all fragmented like irregular pixels on a giant's computer screen. It wasn't until a few minutes later, when we were across the room, that I looked back to that same canvas. Those blobs of color coalesced into a beautiful view of the arched entrance to the Rouen Cathedral. I think time is like that, like your scrim. It smoothens things out, lets them simmer together so that when you get some distance from the past, it turns into something that makes sense. But if you play with those brush strokes…" He closed his eyes without finishing his sentence, but he didn't need to. He opened his eyes and locked them on her. "Kat, you can't go on. You can't—"

She was just wondering if "smoothens" was a real word when Ajit's phone rang. He glanced at the screen. "It's Xavier. Why is he calling me?"

Katya patted her pocket. "My phone's still at work. I left it before my slip."

He tapped the screen and brought the phone to his ear. "Joshi here. I see. I'm sorry to hear that…And why would there be any confusion?… I see…Yes, of course." He hung up. He reached across the table and put his hand on Katya's. "Todd didn't make it. He was declared DOA at the hospital. They're saying it was a drug overdose, but there's some uncertainty. I'm so sorry."

"Didn't make it?" she repeated quietly. She put her head in her hands. Her father let her mourn.

62

Things Are a Little Off

As soon as Brad's flight touched down at Denver International, he tapped his phone off of airplane mode and texted Katya again.

DID YOU GET MY MESSAGE? ARE YOU OK?

As before, there was no response. He rented a car, screamed out of the lot and merged onto I-70. He would make it to Fort Collins within the hour.

Maybe her phone was dead. Maybe she lost her phone. Maybe she missed her flight. Maybe she was sprawled out in some hospital because her cabbie wasn't careful on the way to the airport. He felt disquieted, uneasy, for no objective reason. Lots of people didn't answer their phones.

Maybe she was mad at him. No, that was crazy. She probably wasn't giving him a second thought. But he knew she could be in danger from van Dijk or her ex-husband. What bothered him was that he might be the only one who knew, which made him the only one who could save her from whatever danger might be coming her way.

* * *

A stifling melancholy had settled upon the office. Xavier kept the meeting short, outlining details of his vision for the future of ChronoCorp. After a moment of silence in honor of Todd, he dismissed everyone. The conference room was quiet now. He scraped his eyes with the backs of his hands, yawned, and leaned back in the leather chair at the head of the long table.

A light rain pattered against the one panoramic window at the far end of the room. He'd put the coffee pot on to brew, and its grumbling was the only other sound in the conference room, a sound a lot like one would expect from

© The Author(s), under exclusive license to Springer Nature Switzerland AG 2021
M. Carroll, *Plato's Labyrinth*, Science and Fiction,
https://doi.org/10.1007/978-3-030-91709-8_62

Frankenstein's laboratory. The headaches still dogged him, and the voices still haunted him, but he had discovered enough to solidify his plan, and his doctor had called in something stronger for the headaches. His plan was firmly in place. The first move in his grand scheme had fallen flat, but it wasn't technically him that botched the attempt on Devereaux. The chances of success for the next moves were much brighter. He would need to arrange for the disappearance of several more people, cogs messing up the timeline.

Katya came through his door. "I thought you went home," he snapped. "It's been a long day for all of us. You should go get some sleep."

She didn't sit. Her manner disturbed him. She didn't seem to be mourning the loss of Todd. Something had shifted in her mood since she had left the meeting minutes ago. She appeared taciturn. She walked the length of the table toward Xavier, brooding. "You and I lost a very good friend tonight."

"Yes, we did."

She stopped as she reached the end of the polished tabletop, and stood looking down at him. "I've been thinking," she said.

"Sounds dangerous."

She said nothing for a moment, watching the patterns of raindrops slalom down the window glass. "What if the assassination attempt during our conference was carried out by someone from another time? A time traveler who used our technology and is now trying to undo it?"

Kat was too smart for her own good. Xavier said, "Far-fetched."

"You've got to admit it's a possibility."

A shiver coursed through Xavier's core. Suddenly, it was as if he was waking up, getting his bearings in reality. He had been at the conference, vaguely aware of his plan unfolding at Devereaux's Michigan home. The symposium was his alibi. But he also knew that he had been in the other place, too. A version of himself had, thanks to Emil's multiple universe equations. This other version of himself had been drifting in and out of focus over the past days, just as other memories had. Now, he could see it clearly, and he understood that another Xavier Stengel was out there, up at the cabin. That was now a problem. He wondered if it was a crime to kill oneself. It was something to worry about later, after he gave Xavier Ω his final assignment: Mila van Dijk and perhaps a few others.

Katya interrupted his thoughts. "It's time we stopped all this. And the only responsible way to do that is to do something about the Primus Imperium. After we disassemble our Transport Arena, of course."

"Oh, of course." Xavier kept the sarcasm from his voice and pressed his hands together with elbows on the table, as if praying. "Did you have something in mind?"

She turned away and began to walk back toward the window across the room. "I've been over it and over it. If Rex is right and the Imperium is moving fast, we need to make our move right away. I can't think of any legal way to shut them down before they start doing real damage."

Good, Xavier thought. *We are finally on the same page.*

She didn't look at him. She studied the people outside, three floors below, retreating under their umbrellas or running beneath brief cases or magazines held above their heads.

"Well, I think it's pretty obvious," she said. "If we want to do it in the most efficient manner, we need to yank the Imperium out by its roots. Burn their work to the ground—literally. Ideally, we do it without getting caught and without hurting anyone. And then do the same with ours. It's the only way."

Xavier felt his breath catch in his throat. "This isn't some misguided tribute to our fallen colleague, is it?" Xavier scoffed. "Destroying our work won't bring Todd back, you know."

"It's not about bringing Todd back. It's about undoing the mess we've made."

"There is another way," Xavier said. "We've worked so long and so hard on this technology to just throw it away. Instead, why not make it work for us, not against us?"

"You're sounding very Dickensian," she said.

"Dickensian? Like Scrooge?"

"You're up on your English lit! In the Christmas Carol, Scrooge says something like, 'Men's courses will foreshadow certain ends, but if those courses be departed from, the ends will change.' Is that what you want? Going back to rewrite our own courses so the end changes?"

"All we would need is minor changes."

"Minor changes like what?" Katya asked. "A change in someone's personal schedule here? The murder of one or two key players there, like somebody tried with Arvin Devereaux? How far will this thing multiply? I think you want to go after people, not facilities. You've got this tiger by the tail and you can't let go, can you?"

"You make it sound so Damoclean," he said. "But when all is said and done, no one will know we did anything. Life will return to normal to all but a few people who know the difference. Who knows? We might not even remember what happened."

Katya scowled. "It was time travel that got us into this mess. And it's making us sick. We should avoid using it to get us out. We need to operate within the rules of the universe as we know it now."

"Just how would you do that?"

"We destroy the Primus Imperium's facilities and their files—not the people who made them. You said Emil has made some kind of virus or worm or what-have-you to get rid of their archives. After that, we get rid of the technologies here, save what we can of our research pertaining to archaeology and quantum physics and cosmology, but essentially erase our—the world's—capacity for time travel."

"Because that sounds so easy without time travel," he grumbled. "Let's see, step one: assemble an army. And what was step two again?"

"No, no, no. We can do it without a mob. In fact, we can do it with a handful of strategically placed people. Burn their quantum facility down, making sure to dismantle a few key pieces of hardware, and Bob's your uncle."

"And what if they've brought in a bunch of spear-toting thugs already?"

"That's why time is of the essence, so to speak."

Xavier lanced her with a skeptical glare. "I think we need to consider this carefully before we do anything rash."

"Every event in the past few weeks has fallen under the 'rash' category: shifting realities, past catastrophes, not to mention those spear-toters."

"My dear, this is a battle you will not win today. Let's revisit the discussion when we've both had some sleep."

Katya looked at Xavier with fondness. Yes, he was the boss, but they had shared some of life's most important events. She reached over and patted him on the shoulder with a disarming smile. "Of all the people in this office, you are still the most stubborn. I'm beat. I'm going home."

Xavier wished Katya a good night. As she left, his vision clouded with outright panic. It wasn't the technology that must be stopped; it was the people behind it. Desperate times call for desperate measures, and yet she wanted to play everything so carefully, so conservatively. It just wouldn't do. He must concentrate on the players, not the stage.

Fortunately, he didn't need Katya for his plan of salvation for the universe. He didn't really need anyone except for his accomplice, who would be along again soon.

How many Xaviers were there, worrying about their own universes at this moment? If every person represented some sort of bifurcation, some new reality, whose universe was he in now? Was it Bradley Glenn's, the wonderkid from the museum? Was he wallowing in the reality governed by Todd or Katya? Was he now—heaven forbid—locked in the universe of the Ambassador? It was all too much to get his head around. His hands began to quake. If there were these multiverses, every person's decision every hour of every day gave rise to a new reality. But surely there must be a limit. Did a universe split off when Xavier used his turn signal or not? When he sent an email on a Monday rather than a Tuesday?

He began to tremble. Perhaps one action made enough of an impression in the timeline to stick, while another didn't have staying power. Still, the possibilities were humbling. It was all so wondrous, and so terrifying. Did he really have that power? Perhaps that's where a higher power came in, but Xavier simply couldn't make room for something like Todd's God. The possibilities of time travel were too grand, the stakes too extraordinary, to simply yield to fate. No: he had to try to do his best in this universe, this reality, this moment. His father's words echoed in his mind: *If it is to be, it is up to me.*

The universe was a mess, and he was the only one who could put it back the way it was designed to be. Kat was right. It had to be done. Or undone.

63

The End of a Long Day

Katya dragged herself into the elevator, punched P1, and grabbed her phone, freshly charged from her desk. She couldn't bring herself to call Mila, although she had promised to as soon as she was back. The loss of Todd had drained her emotionally and physically. The cliché of someone's departure leaving a hole or a vacuum didn't seem to fit. Katya felt more like one of her limbs had been amputated, the loss necessitating an entire relearning of how to live each day.

She leaned against the elevator wall. Her thoughts drifted to Brad. When she thought about him, she could feel that electric current flowing, and she knew it was flowing between the two of them, rather than making a dead-end circuit. She saw it in his eyes at dinner. She heard it in his voice when he encouraged her.

What was she thinking? She wondered as she walked out of the building. Love. Bah. This attraction was nothing but hormones, a bunch of juices flowing through the veins, short-circuiting the thought processes. It was that irrepressible primeval drive, hard-wired into some Darwinian strategy, spurring on that impulse that would pass down the genes. It was all so ridiculous. She was a scientist, for God's sake. Rational thinking formed the backbone of everything she did.

But she took a breath. Love was not a bunch of juices and primal programming, any more than the Taj Mahal was a bunch of bricks. No matter how one tried to dissect it, love was greater than the sum of its parts.

Her luggage from the conference was still down in the trunk of her car, waiting in the parking garage across the street. The lights of ChronoCorp reflected on the rain-bathed pavement. The crimsons and golds of sunset were draining away, replaced by twilight's purple and pink. Lightning flared in the

M. Carroll, *Plato's Labyrinth*, Science and Fiction,
https://doi.org/10.1007/978-3-030-91709-8_63

distance, but the clouds overhead had torn apart to reveal the stars. Passing headlights added to the fluid, incandescent shimmer of the wet pavement, like multicolored lava.

Her mind raced. When she had patted Xavier's shoulder, it was partly to see his reaction. Strangely, he had none. She expected some sort of friendly acknowledgement, some reciprocation, but it was as if the old Xavier was gone and a new one was taking over, inch by inch. Poor, brilliant Xavier. He was the perfect portrait of humanity, a species that had learned to live long, but not necessarily smart. Perhaps the human race as a whole had advanced in intelligence, but the application of knowledge—wisdom—did not always follow. Todd had always suggested that we were called to rise above that nature. Could we? She wondered.

At present, it seemed as though their work had triggered some kind of imbalance, as if things were on the brink of spiraling out of control. Xavier was about to do something big; she could feel it. She had to move now, and she would need help.

As she approached the ground level, several messages from Brad scrolled across her phone. Yes, Brad would be her ally. Perhaps even Rex, too. She swiped her screen, but before she could read Brad's notes, a cherry-red Miata pulled up, its roof stowed behind the two seats. The driver wore tinted glasses and a scarf around her head despite the late hour.

The woman slid her shades down the bridge of her nose and said, "Hello, my dear. Can we talk? Get in." It was Mila van Dijk. This was shaping up to be a painfully long night.

How did Mila know I would be here? What was she doing in town, driving herself?

As if reading her mind, Mila said, "I was dropping off a friend who flew in from New York. My literary agent, in fact. Did I tell you I'm working on several books about antiquities?"

"No, you didn't." Katya climbed in and carefully shut the polished door. "I had no idea you were a writer. Is it hard?"

Mila pulled her scarf off and let the wind blow her hair as she drove. "Do you know what Jean Shepherd said about the subject? He said, 'Writing is easy. All you do is stare at a blank sheet of paper until drops of blood form on your forehead.'"

"Was he right?" Katya asked.

"Pretty much."

Mila's hot little car climbed its way into the foothills outside of Fort Collins. The mansion loomed ahead, its glass reflecting a nearly full moon. As they passed the manicured lawns and hedge sculptures leading up to the driveway,

Katya spotted the hunched figure of the gardener digging beside one of the garden lamps, still at work in the twilight. She wondered if he ever stopped.

As Mila ushered Katya into the main hall, Katya sensed that something was wrong. The ancient statuary was still there, and the Calder still swung from the vaulted ceiling. But the place seemed to have an edge to it. Electricity charged the atmosphere. Mila had become nervous, and the silent mansion did not smack of celebration. Whatever Mila's news was, it wasn't going to be good.

Mila walked over to the wall and pressed a button. "David we're here now."

"David?" Katya asked.

"He is a nice man, someone who will see to it that no harm comes to you."

Katya fought down the panic in her chest. "What do you mean, harm?"

Mila didn't answer, but turned toward the window, looking crestfallen. Two very large men stepped into the room.

"This way, miss."

Katya took one last look at Mila as the two burly oafs guided her toward the door. Mila's eyes were brimming but she would not meet Katya's gaze. "Mila…" she began. But as she reflected on the last few weeks, she understood all too well. All this time, their benefactor had been spying on them. And now, to put a sharp point to it, Katya was being kidnapped.

* * *

Brad slammed the door on his rental and dashed toward the ChronoCorp entrance at a dead run. He slid through the front door and headed into the elevator.

The doors opened onto the third floor lobby of ChronoCorp. A weary receptionist was still stationed at her desk.

"May I help you?" she asked.

"Katya Joshi," he said, trying for a friendly smile.

"I'm sorry; she's not here. Her schedule shows her coming in tomorrow morning at eight."

"One can only hope," he said under his breath. To the receptionist he said, "What about Xavier?"

She gestured down the hallway. "Conference room. Third door on the left. And could you please remind him that I'm still here?"

Brad found Xavier huddled over the table, back toward the door. Even from behind, Brad could tell the man was not at his best. His hair was disheveled, shoulders slumped. His gaze seemed focused on some empty spot on the table.

"Xavier?" Brad said tentatively.

Xavier straightened in his chair and swiveled to face him. "Well, well, twice in one day. I can only guess that you've heard about Todd."

Xavier's face was careworn, his eyes bloodshot. Though he hadn't been sure during the whirlwind conference, now, Brad could see why Katya was concerned for the man's mental health.

"Todd?" Brad asked. "It's Katya I was worried about. What about Todd?"

"I'm afraid we lost him."

"I'm happy to help out; go back to ancient Greece and search."

"No, Brad. He died. Just after his last slip."

"Died? I'm so sorry."

"Time travel seems to be hard on the body," Xavier said morosely.

"Any idea why we lost him?" Brad asked, using Xavier's language.

"We're looking into it."

"I'm so sorry," Brad said again.

Xavier nodded, still focused on the tabletop. "But that's not why you made your detour to our fair Rocky Mountain town. What's on our mind?"

"It's about Katya. I don't suppose you have any idea where she's gone off to?"

"Sometimes this place makes me feel like a mother hen." Xavier squinted his eyes shut for a few moments, let out a wheezy sigh, and said, "Our principal donor did ask when our team would return from San Francisco. She sounded a bit urgent about it, but I assumed Katya was heading home for the night before she went anywhere else."

Brad felt the tension building in his chest. "If you don't mind me asking, how much does Mila van Dijk know about what transpired at the conference?"

A look of surprise flickered across Xavier's face. "Ah, you know of our funding source. We'll be putting together a summary report for her tomorrow, once everyone has a good night's sleep."

Brad pulled out the chair next to Xavier, sat down and met his eyes. "I think Mila may be more than she appears. I suspect the guy running the show at the Primus Imperium is her ex-husband. A real stand-up kind of guy. Did a stint in the pen for arms deals. Who knows what else he's gotten himself into?"

"And you think he and Mila are still in contact?" Xavier scoffed. "That would mean Mila is carrying out some kind of industrial espionage. I'm having a hard time seeing it, especially seeing how she is funding the majority of our work. Besides, we vetted her. Aside from a fixation on really old stuff, she is perfectly normal."

"Maybe," Brad said. "But maybe not. She's got endless money. How do you know Mila is even her real name?"

Xavier remained silent, but his look of panic said it all.

"Maybe the Imperium guy has something to hold over her head," Brad offered. "Maybe they're still involved romantically or something. But I know one thing: Mila asked Katya to meet her as soon as she arrived here, and now Kat's not answering her phone."

Xavier frowned and held his chin. "Not like her."

"I know, right? I think she may be in real trouble." Brad stood and paced nervously to the window, his hands fisted at the small of his back. "Where would Mila take her?"

Xavier muttered, "I was just thinking of Minoan pots and rows of sphinxes."

"Sphinxes?"

Xavier stood, holding his forehead. "I think I know where she might be."

Brad said, "I'll drive."

64

Xavier and the Ambassador

It was one of the most surprising phone calls Xavier Stengel had ever received. As the road began to wind into the foothills, Xavier's cell went off. The caller I.D. was blocked, but he was curious enough to take it anyway.

"Stengel here."

"Dr. Stengel, I have been watching the progress of ChronoCorp with great interest. I congratulate you on your recent work."

"Thank you," Xavier said suspiciously. He glanced at Brad, whose eyes were glued to the road. "What, specifically, interests you?"

"Some of your less public work. For example, the bioimaging system than can record audio as well as video. Clever."

Xavier froze. Emil's new audio technology had supposedly remained a company secret. How much did the caller know? "Can you be more specific?" Xavier asked. "I'm not sure, precisely, what you mean."

After a pause, the man on the other end said, "Dr. Stengel. I have very little time and even less patience for cat-and-mouse games. I will tell you that I, too, am working with a group of visionaries involved in temporal research. I'm contacting you in hopes that we can not only compare notes, but perhaps even combine forces."

"For someone who has no time for games, you haven't been very forthright about in your own identity."

"Fair enough. I am the head of an organization that has been painted with a wide brush of misinformation. It is called the Primus Imperium. Before you decide that we are a bunch of terrorists or any such nonsense, may I share our vision with you? I think you will find we have much in common."

"And your name?"

M. Carroll, *Plato's Labyrinth*, Science and Fiction,
https://doi.org/10.1007/978-3-030-91709-8_64

"Bouchet. Louis Bouchet, but my colleagues know me as the Ambassador."

Xavier had visions of Rex bumbling into ChronoCorp, committing industrial espionage at Bouchet's bidding. "I will tell you, Mr. Ambassador, that I am not inclined to work with someone who has been spying on my organization."

"We can bring those days to an end, you and I. To paraphrase a cliché, we are stronger together than working toward separate goals."

"That would be true if we shared the same vision for the end game. Just what are you offering?"

"A chance to be in on the ground floor of a new world, a new way of seeing reality, a political system ruled by peacekeepers rather than combat and subterfuge. I would bet that is close to your vision of the future as well, and so I come to you."

"I see," Xavier said. He felt moisture on his palms. "I'll bet you're running scared because of our plans to cease research."

"Dr. Stengel, please. We both know that is not going to happen, with your organization or ours. You have your Transport Arena with your time pods. We have our Transference Theater and time capsules. You say tomato. Except that we've come up with something a bit more advanced. I call it a chronometric coach."

"Has a nice ring to it."

"It should," the Ambassador said proudly. "It can carry twenty people—heavy infantry—at one time."

Xavier fought a feverish shudder. "Infantry? You're building troop carriers?"

"Yes, and it's brilliant," the Ambassador gushed. "We're plucking our soldiers from the night before a disastrous battle—disastrous for them, in fact. How's your Roman history?"

"Rusty."

"Well let me tell you, Dr. Stengel: the Battle of the Teutoburg Forest wiped out most of three Roman legions in 9 AD. It was a massacre, and they knew it would be. We'll be giving them the opportunity to miss that suicide, a chance at a new life in a new time."

Xavier's head was throbbing and his pulse was racing. Normally he could muster up a congenial tone, but he just didn't have it in him. He said, "You know what I think? I think you are one deranged puppy. Thanks for the call." He hung up.

"That was…colorful," Brad said.

"And informative. Tell you about it later." Xavier pointed up the road. "It's just up here on the right."

Brad was looking at the parade of sculpted shrubbery and nearly missed the wide driveway. The two men climbed out of Brad's rental and made their way through the gauntlet of Egyptian statues.

Xavier was again impressed with the massive front entry. He pulled a bell cord. Bells tolled behind the glassed front walls. In moments, a butler opened the front door. He lifted an eyebrow.

"May I help you?"

"Xavier Stengel and Bradley Glenn to see Ms. Van Dijk?" Xavier said uncertainly.

"One moment, sir. Please, step inside." The butler gestured for them to enter the expansive piano nobile. Brad whistled. The butler returned almost immediately.

"I'm afraid the lady of the house is not in residence at the moment."

Xavier glanced at Brad, then said, "Do you know where she went?"

"I didn't feel it was my place to ask, sir."

"No," Xavier fidgeted. "Of course not. I don't suppose you know when she'll be back."

"Sorry, sir," the butler said as if he had already answered the question.

"Thank you," Xavier said, turning back toward the car.

"Would you like to leave a message of any kind, sir?"

"Just tell her—just tell her—" But what was there to say? "No message, thanks."

Xavier slammed his door. Brad pulled away, obviously watching the rows of sphinxes and topiaries parading by. He pulled his eyes back to the road and asked Xavier, "Do you think Mila van Dijk is really not there?"

"We have history. I'd like to think we have some degree of trust, at least. I think if she knew I was here, she would come out to talk to us."

"I guess. But what about Kat? Even if van Dijk's not here, they might be holding Kat somewhere in that big MacMansion."

"If anyone is holding Kat against her will, it's not going to be Mila. My money's on the Imperium warehouse."

Brad said, "Do you have a way of getting in touch with Rex?"

"I do. Why?"

"I think it's about time we put his allegiance to the test. I just hope he passes. We need him. I'm thinking they took Kat somewhere out of town."

"Do you think they crossed the border into Wyoming?"

"If I was them, I wouldn't take the time. I think she's somewhere closer to home."

"Home being the HQ of the Primus Imperium," said Xavier. "There's plenty of room in that building to keep an extra person."

Brad nodded and dropped the car into a lower gear. "Guess we'll just have to go find out."

Xavier pulled out his phone. "I need to get Emil to work on that worm for the Imperium's software. He's already got files from Rex's last visit, and he thinks they'll be enough. And there's something else I've got to do. You get Rex and see if you can find Kat at the Imperium. Emil also may need help getting into the Imperium headquarters to set the worm. Drop me off on the way, and I'll be along as soon as I can." Xavier was seeing spots, and his headache was back. He was having a hard time focusing on the road ahead.

"Must be important, if it can't wait for the rescue of your colleague."

"More than you know," Xavier said.

65

Pink Elephants and P.I.s

"Are you off your little Dutch rocker?" Rex bellowed into the phone. "How did you get this number, anyway?"

"Xavier," Brad said.

"I told him not to give it out."

"Katya's been kidnapped. By the Ambassador's people, apparently. Xavier made an exception."

Rex grumbled something unintelligible, then fell silent for a few moments. "Okay, if I know the Ambassador—which I don't—I'm guessing he'll keep her close. She's obviously a bargaining chip, so he values her. And yeah, they have some CCTV around, but the place is remarkably light on guards. We can distract them."

"Distract them how?"

"Some well-timed arson."

"You mean burn the place down?"

"Nothing so dramatic. Just start a modest conflagration to shift the Ambassador's attention while we look for Dr. Joshi."

"What could possibly go wrong?" Brad mocked.

"I'm a man of many talents," Rex said. "Strategic fires are something I'm pretty good at. I've had practice. In a past life, I'd had a pretty lucrative business involving arson at well-placed sites… most of them heavily insured. There's another problem, though. They've got a second facility. I found references to it in some Imperium files. It's up the Poudre Canyon, and it's some sort of armory. That place is probably crawling with guards and scientists and who-knows-what-else. If they have Katya there, our distraction at the

M. Carroll, *Plato's Labyrinth*, Science and Fiction,
https://doi.org/10.1007/978-3-030-91709-8_65

warehouse won't help. A little indoor camp fire at their Poudre facility at about the same instant just might do the trick."

"It's a lot harder to handle two crises at once," Brad agreed. "That will keep them busy, but we'll need more help."

"I guess you and Xavier try to find Kat at the warehouse while I take care of the place up north."

"So let me get this straight," Brad said. "Xavier and I go to the facility here in town and wait for you to create a minor catastrophe up at the other place, at which time we light our fire and search for Kat and hope we find her in the three minutes that the guards are distracted?"

"I can give you more than three minutes. The first thing we do is disable the fire suppression system so the little fires we set won't go out right away. I'll draw you a diagram of how to fix the sprinklers at the warehouse. Then, we set off the alarms so everybody gets out. Fire alarms in both places ought to get their attention. Between the two events, I'm sure you'll have a few minutes to look around."

"There's another thing on our list: Emil has to get that snail or worm or whatever it is into their computer system."

"Okay, so Emil goes along."

"And who goes with you?"

"I can work alone. I've done it before. I'll need a few hours to get things together; I don't keep C-4 lying around the apartment. Maybe shoot for cover of darkness, the night shift tomorrow?"

Brad sighed. "In the what-could-possibly-go-wrong department, this has them all beat."

"It's all I got," Rex said. "And unless you've got something more solid, that's the plan."

Brad had to admit it: he had nothing more solid.

66

Exit Strategy

The Ambassador informed Dresden that he was sending a car for him. It was less an announcement than a command, and Dresden wasn't about to argue. The SUV that pulled up was familiar: it was the same vehicle that had taken him to the north building earlier. Inside, the same two guys wearing the same suits and—he assumed—packing the same firearms, greeted him.

"So," Dresden said in a friendly tone. "It's back to the North Facility?"

"Apparently," the driver said, pulling into traffic and heading for the Poudre Canyon. In minutes, they were coasting through the familiar parking lot of the large, columned citadel bedecked with banners of world empires. As the vehicle came to a stop, the driver said, "The Ambassador wants you to see how his plans are progressing. He'll be here in a while to meet you, but for now—"

He gestured toward two very large guards who approached silently. One guard was a woman, compact and frowning. She looked as fit and capable as the man beside her. Both wore nametags, but without the friendly "Hello, my name is…" at the top. The man, Clive, said, "Follow us, please."

Dresden fell in behind the man while the woman marched behind. He felt more prisoner than guest, but that was probably the Ambassador's intention. They passed through the entrance, went through the same formalities that Dresden had endured before, and shuffled to a brightly lit side room about the size of a jail cell.

"Please wait here," the woman said.

He did, for a good twenty minutes. When they returned, the male guard had an interesting announcement. "The Ambassador will be calling you on your cell momentarily."

The guards stepped out but left the door open. Dresden's phone rang.

© The Author(s), under exclusive license to Springer Nature Switzerland AG 2021
M. Carroll, *Plato's Labyrinth*, Science and Fiction,
https://doi.org/10.1007/978-3-030-91709-8_66

"Dresden, I got held up. Why don't you go on, have a look around and feast your eyes. It will inspire you. I've spoken to the guards. They'll admit you and Clive will show you the way."

"All right," Dresden said, seeing no other option.

The brute named Clive escorted him down the white halls and through the double doors. They stood at the edge of the balcony he had seen before, the one that ringed the mess hall. But the vast room was no longer quiet. In fact, it sounded like a controlled riot. A banquet blanketed the tables twenty feet below: platters piled high with meats, trays of freshly baked bread, tubs filled with grapes and melon. Suddenly, he was on the shores of Elysium, gazing upon a sea of Apollos and Adonises. They laughed and guffawed and blubbered around mouthfuls of food, row upon row of Roman soldiers, a freshly minted phalanx from Michelangelo's pristine marbles. Some wore swarthy dark brows and coal-black eyes, while others favored curled blonde heads of hair. The older troopers leaned toward balding and gray, but their bodies remained athletic.

The soldiers represented the Olympian ideal. But at the same time, there was something unseemly, almost hideous, about the scene. The warriors were completely out of place, like a graceful sea serpent yanked from the depths and beached on the desolate shoreline to flop around aimlessly.

Dresden scanned the great room, a place with all the charm of a cut-rate gymnasium fused with a fast-food establishment. The bench that ran along the back wall was no longer empty. Spread across it lay an assortment of weapons in various stages of disassembly. Someone had been educating these Roman shield-bearers in the ways of automatic weapons and RPGs. The countertop preparations looked quite unlike those of a peacekeeping force.

At the far corner, two Romans faced each other at about five paces. One appeared to be taking the other's photo on a cell phone. Then he tossed the phone to the other. A guard came over, had the men stand together, and showed them how to take a selfie.

"Oh yes," Dresden said to the guard. "Modern technology is going to be great for these guys."

So, the Ambassador had finally done it. His plans were unfolding as designed in their complete and nightmarish form. Dresden began to think seriously about the details of his exit. If he had been married, tied down to a family, or if he had a bunch of horses or tracts of real estate, it would be difficult to cut his ties quickly. But he had none of those things. He was free and unencumbered. All he had to do was toss his few worldly items into his truck, and take off to anywhere he desired. The only passenger he had to concern himself with would be Tobie, his little Chihuahua. Despite her tiny teeth—like a piranha, according to one tech—the diminutive Chihuahua was a fierce canine. And odd. Dresden often caught her barking and snarling at her own left rear paw, a look of paranoid panic in her eyes. She was definitely broken. So broken. The people at the rescue weren't even sure she could be rehabilitated. But they prescribed a drug regimen, gave her some rudimentary training, changed her name from Fang to Mimi and pushed her up for adoption on a wing and a prayer.

When Dresden had adopted the little beast from the shelter—after all, he reasoned, aren't we all broken in some way?—Mimi still wore her collar reading "Fang." Dresden had re-christened the puppy with the more innocuous name of Tobie. On a routine veterinary visit, Tobie had used her toothpick teeth to inflict seven stitches' worth of damage on an aggressive and overconfident Rottweiler. Dresden was fond of the miniatures for the same reasons that others despised them: they could carry a surprising amount of ferocity in an unexpected package.

Dresden's handler pulled his wrist to his mouth. He squinted and, with his other hand, shoved his earpiece in further, as if to change whatever message was coming in.

"Now?" he whined. "I'm a little busy. Yes, I know that, but I have a charge that I'm watching—yes, him. The Ambassador? I see. Be right there."

Clive turned to Dresden. "It seems I am needed," he said with acid in his voice. "Please stay on this balcony or in the adjoining room. There's coffee. I or another guard will be back soon."

"Will do," Dresden said. But turning back to the scene before him, he knew it was definitely time to hit the road. He'd made more than adequate preparations. He would leave. Tonight. As soon as these goons got him back home.

He nodded, satisfied with his decision. But looking down, he saw something that made him realize his work here was not yet done.

* * *

Bradley Glenn checked into a motel on the south side of town. He had watched as Xavier left ChronoCorp the night before. He wondered where the man could possibly be going, but Xavier had been evasive. Brad got little sleep, showered and inhaled breakfast, and returned to a bustling ChronoCorp.

"Good morning," he said to the receptionist. "Is Xavier in yet?"

"He's late this morning. You're welcome to wait in the lounge."

He walked down the hall and settled himself in the employee lounge. He selected a famous tube-shaped snack from a machine and took a bite. White cream oozed onto his cheek. Emil came in and smiled.

"You're making a face, Dr. Glenn. Something wrong with your gourmet snack?"

"I really thought these things had too many chemicals to go stale."

"That one's probably been in the dispenser since the early Pleistocene. I've finished."

"Finished?"

Emil held up a flash drive. "The worm. It will make its way past just about any system security and corrupt all the base files. Then it goes to work erasing all the files stored in the system. It's polymorphic and multipartite."

"Poly what?" Brad asked.

"Polymorphic. Think of a hydra. It tries to get into the system in lots of places. Even though the Imperium's security software will try to shunt it off from other files, it just keeps hammering away until it finds a new route, erasing things as it goes. I'm very proud of this baby. It takes care of route directories, archives, the whole shooting match. All you do is plug it into any port in their office. I'll have to infiltrate the cloud itself with a separate virus. That's not easy."

Emil tried to hand the flash drive to Brad, but Brad stuffed his hands behind his back and shook his head. "I'm lousy with technology."

"This thing is plug and play."

"Emil, I need help. I haven't the slightest clue where they've got Kat, and from what I hear, it's a big building. Can you come along? You can install that thing," Brad gestured toward the little flash drive, "and then we can both look for Kat."

Emil looked at the ceiling. "Sounds like a grand outing. Actually, it sounds like we need a battalion with us."

"I'll go," came a voice through the door. Brianne stepped in. "Katya's been good to me. So has Emil, and I don't want anything happening to him or the rest of you. Happy to tag along. Besides, I have a brown belt in Aikido."

Emil raised his eyebrows. "You're just full of surprises, aren't you?"

"I have a brown belt, too," Brad grinned. "It's from Neiman Marcus. We appreciate your help."

"I want to get something from the Transport Arena that might be useful," Brianne said. She was gone for only moments, and returned with the shovel. Brad found himself reading the words again: *Sometimes it just piles up dark and deep. That's when you start shoveling.*

Brianne looked sheepish. "Even with martial arts," she said, "sometimes a little brute force is called for."

"I guess it's time for us to start shoveling," Brad said.

The trio went to the front desk. "Any word from the boss?" Emil asked the receptionist. She shook her head.

Brad imagined Kat in some dark, rat-infested room, chained to a wall, duct tape over her mouth. "I hate the idea of leaving Kat there, but if we went now, we'd be going at the busiest time."

Emil held up the little flash drive. "If we're going to 'save the universe,' as Xavier put it, the sooner I set this thing to work, the better."

"And if we get caught," Brianne said, "you won't get to try it out."

"I suppose you guys are right. Tonight, then?"

Emil nodded. "We'll have to coordinate with Rex. Hopefully Xavier will show by then. Where is he, anyway?"

*   *   *

The drive up Poudre Canyon usually relaxed Xavier, but circumstances had wound him up. The stress of hopping through the ages had left him coiled like a spring, and the thought of removing people from the timeline didn't help.

He slowed the car and turned onto a rutted road, a passage that served as the driveway to his mountain getaway. The cabin nestled in a stand of evergreens, a ramshackle collection of boards clinging to a stone chimney. Parking a few yards down the hill, Xavier stepped out with an armload of his favorite foods: graham crackers, cheese pops, and dark chocolate. He assumed that Xavier Ω would feel the same way.

Xavier shoved the front door. It swung open with a loud creak. Xavier Ω looked up from a magazine lying on the table with an expression of surprise.

"Well, look what the…"

"Cat dragged in," Xavier finished for him. "I know, I know. I brought you some more sustenance."

Xavier Ω practically slobbered over the treats. "Very nice. Just what the doctor ordered."

"Yes, well, speaking of which," Xavier said tentatively, "I have another favor to ask."

"Anybody who brings me cheese pops deserves at least one favor."

Xavier felt his pulse race. At the same time, a profound feeling of sadness blanketed him. "It's a very hard thing to ask," he said.

"Shoot," Xavier Ω said casually.

"Actually, you're the one who will do the shooting. I need you to assassinate Mila van Dijk." Xavier's voice cracked as he said it.

"So, there seem to be some things missing between our two universes," Xavier Ω cut in on his thoughts. "Just who is Mila van Dijk?"

Xavier couldn't believe it. "Mila van Dijk is the philanthropist funding our operations at ChronoCorp. I assume you remember ChronoCorp?"

Xavier Ω scowled. "Don't be daft. Of course. So you're saying this van Dijk character is funding ChronoCorp…but does so privately, and isn't in fact your Secretary of Defense?"

"Defense?" Xavier howled. "Do you mean to tell me the ChronoCorp in your universe is a government entity?"

"Yes, but apparently, all the red tape that comes with federal funding means we've moved much slower than you. Where does this Mila live?"

"I will have a map for you, just like last time," Xavier said. He laid a Glock on the table. "Be careful with this. It's the only one I have that actually works. But can you please aim more carefully this time? Your last target is alive and well."

Xavier Ω tapped the newspaper on the table. "So I hear. Hey, can we get internet up here?"

"Sorry," Xavier said, tossing the morning's newspaper on to the table. "I'll take you to a rental car agency soon, and you'll go from there. Just wait for me. It will be within the next day or so."

67

Electronic Escapades

Bradley parked in the alley near the Imperium headquarters. He had waited as long as he could for Xavier, but the clock was ticking, Rex would be on his way up the Poudre Canyon any time now, and the night shift was on. Emil and Brianne pulled up behind him in Brianne's car. Brad stepped from his car carrying a small paper bag. Emil led them to the end of the alley, where they studied the warehouse. Wind moaned through the broken glass of the dark windows. No light shone out, but faint sounds emanated from somewhere inside. The bricks and siding had pulled apart, exposing the metal and timber skeleton of the edifice.

"You sure this is the right place?" Brianne asked Emil.

"Xavier gave us specific instructions. The best door to go through is that one over there."

Brad raised an eyebrow skeptically. "Okay. First order of business is to find Emil an outlet for his little magical device."

The door had an electronic lock, but Emil had a clever key to decode it. He leaned over to study the mechanism.

The entire entry area was bathed in light from above. Brad felt like a sitting duck. He turned to Brianne, who was gazing at the overhead bulb.

"Can I give you a boost?" he said. He cradled his hands. Brianne stepped up and unscrewed the bulb. Under cover of darkness, the door clicked and Emil stood up. "Should I leave it open a bit in case Xavier shows?"

"Good thinking," Brad said.

In moments, the three were creeping down the dimly lit hallway. After finding several closets and small offices, they stumbled upon a larger room with a complex computer outlet, equipment lining the walls. The room was

M. Carroll, *Plato's Labyrinth*, Science and Fiction, https://doi.org/10.1007/978-3-030-91709-8_67

cold, refrigerated to keep all the electronics from overheating. Brianne stepped through the door, leaned the shovel against the wall, and walked to a desk. It supported a computer tower and late-model monitor. Emil went to examine the banks of hardware against the wall. Brad put his paper bag on the computer desk and shoved his hands into his pockets, not knowing what else to do.

"I think someone's been in here with a cat," Emil said.

"Why?" Brianne asked.

Emil pointed to his nose. "Allergies." He sneezed.

"I love cats," Brad said.

"Emil, will this port work?" Brianne asked, pointing at the desktop tower.

Emil tossed the flash drive to Brianne, who caught it with one hand. Brad watched her. She moved the drive toward the computer terminal, but froze, looking beyond Brad.

Brad pivoted to see a guard entering the doorway with a baton. A big guard. *Why do they always have to be so big?* Brad wondered. As the guard entered, a cat followed him in.

"How about if you all don't do anything creative?"

Brianne said, "We're guests of the Ambassador."

"I'm sure you won't mind if I check." He held his wrist up to call for help.

* * *

Xavier parked in the now-familiar alley, behind a car he didn't recognize. Ahead of it was Brad's rental. All three vehicles were cloaked in the darkness of the unlit alley, and Xavier knew that the side of the warehouse up ahead was largely unused. Xavier carried Rex's map, a tangle of lines and boxes scrawled on what may once have been a napkin. An arrow pointed through a gap in a line. Presumably, this was the door Rex had suggested he enter through.

Xavier approached the door. Mercifully, the light above it wasn't working. Using the flashlight on his keychain, he blinked at the map, trying to focus his eyes. He rubbed them, then rubbed the back of his neck. He tried the door. It had been left slightly ajar. Maybe nobody used this entrance.

Xavier entered the long hallway. He looked at the map. There should have been three rooms on the right and then an electrical unit. He followed the corridor until he got to the third door. Instead of an electrical assembly, another hall branched off. He frowned at the map. He was lost.

68

C-4 and Cannelloni

Rex Berringer was in the middle of rigging his own incendiary device when he got one of those phone calls that he always dreaded. He recognized the number immediately. Dresden always called on some secure line associated with the Imperium. The man's quiet drawl came across the line in a chilling timbre.

"Mr. Berringer, I need your help. There is something that I must do, and it will take the two of us to do it right. I will meet you at your office in fifteen minutes. Please do not be late."

This was a truly remarkable event. Dresden had never asked for anything. Dictating was more his style. Something profound must have changed. Rex could probably slip out and skip town before the dark Texan arrived, but now he was conflicted. He didn't want to let his new ChronoCorp friends down, so he had to make it to the North Facility. But just what could Dresden possibly want?

Dresden stepped through Rex's door right on time. He seemed to lack some of the dark energy that usually oozed from his pores. In fact, he seemed nervous.

Rex didn't get up. Savoring the moment, he put his feet up on his desk, leaned back, and said, "Well, Mr. Dresden, what can I do for you?" He hoped Dresden would make it quick. He had to get up the Poudre canyon before Brad and his crew sprang into action at the warehouse.

Dresden tipped the brim of his hat slightly. "They've got Ms. Joshi. The ChronoCorp scientist. Katya."

"Yeah, I know," Rex said evenly.

"I saw her at the North Facility, and she didn't seem to be enjoying herself, what with a blindfold on and hands tied behind her back."

© The Author(s), under exclusive license to Springer Nature Switzerland AG 2021
M. Carroll, *Plato's Labyrinth*, Science and Fiction,
https://doi.org/10.1007/978-3-030-91709-8_68

Rex shot to his feet. "The North Facility!"

Dresden held up his hand. "My advice to you is to not panic. It tends to cloud the thoughts. I have something in mind."

"So do I," Rex said.

* * *

Rex stood beside Dresden, peering over the balcony and smelling aromas of Italian food. He was glad to help Dresden take apart the Imperium, but the North Facility and its people gave him the creeps. And yes, burning things down was something Rex Berringer was good at. But burning down a cafeteria didn't make much sense. From their standpoint on the gallery, the scene was too surreal for Rex to believe.

"It takes food to drive an army," Dresden said. "Just ask Napoleon."

"Is your big plan to destroy all the plasticware?"

"Have a look along the back wall," Dresden drawled. "Those aren't Italian canapés. They're fairly advanced weapons. And besides, it's what's in the basement that counts. The Ambassador has an entire armory, with lots of explosives that should help your combustible operation. Shall we make it all go away?"

Rex scanned the doors behind them. He wasn't sure how Dresden had pulled it off, but the Texan had somehow convinced the Ambassador to let them in. Something about showing Rex around for his future employment with the Imperium. But both Rex and Dresden knew that their relationship with the organization was about to end. Abruptly. "I'm game. But how will we get back out through the gates?"

Dresden smiled, a bit like a crocodile. "The fire alarm automatically opens them so the fire department can get in. I checked. We will have an alarm, yes?"

"Yes, but without the sprinkler system. Everybody will get out but there should be plenty to keep the fire department busy."

Rex's phone vibrated. It was a text from Brad.

HEADED 4 NORTH FAC. MAP?

"Clever plan," Dresden said. "How much time will we have?"

Rex frowned and calculated. "If we're going to actually raze the place to the ground, I'll first have to disable the fire suppression system. Not hard, but it's usually not where I'll need to set the charges. Once those are done, we'll have five or ten minutes before things really get going. No more, if you want to get out less than well done."

"Ten minutes," Dresden said quietly.

"Or five. You can set one charge at the far side while I do the other down-stairs. And you can pull the fire alarm. I'm going to be pretty busy."

"I'll need that full five minutes. I have an errand to run."

"An errand?" Rex said, astounded. "What, do they have a gift shop here?"

"They have something else. Don't forget Katya."

"Never!"

"I'll meet you out front, just beyond the outer fence, across from the Genghis Khan banner. After."

69

Secrets in the Dark

Emil cowered behind the bank of humming electronics against the wall, hidden from the guard's view. Brianne and Brad stood still, watching the guard prepare to make a call that would seal their doom. He felt the sneeze building and there was nothing he could do about it. It was a good, loud one.

The guard said, "Come out where I can keep an eye on you until I get a little help."

A figure entered stealthily through the door behind him. As it came into the light, Brad realized it was Xavier. The boss lifted a gun to the guard's head, chambering a round. "I wouldn't do that if I were you," he said. The guard lowered his radio. Brianne stripped it from his wrist and reached into his pocket for his phone.

"Is there another phone in here?" she asked him.

He nodded toward the back of the room. "On the wall."

She stepped over to the phone and yanked the wire from its outlet. Turning back to the workstation, she said, "I've uploaded our little—" she glanced at the guard, "present." She pulled the drive and handed it to Emil.

Emil said, "Very nice, but now what?" He eyed the husky guard.

"We could tie him up," Xavier offered.

Brianne walked around behind him, grabbed the shovel, and made a quick swing to the back of his head. The guard crumpled like a stack of dominoes.

"I wouldn't want to be on the wrong side of either of you guys," Brad told them.

Xavier smiled and held up the gun. "This is nothing. Doesn't work. Not since I took it through a Q-slip. But you've got to admit, it made a great prop."

"It sure did," Emil agreed.

M. Carroll, *Plato's Labyrinth*, Science and Fiction,
https://doi.org/10.1007/978-3-030-91709-8_69

"I was afraid I couldn't find the place," Xavier said, breathing heavily. "Turns out I was reading Rex's map sideways."

The way Xavier was waving the gun around made Brad nervous. Xavier's demeanor had changed decidedly since their drive to Mila's mansion. His skin was the pallor of sanded pine, accentuated by a sheen of sweat. His bloodshot eyes sank into darkened bowls, and saliva streamed from one side of his mouth. His lips twitched in time to his trembling hands. He seemed on the verge of madness.

"Xavier, we'll find her. Don't worry."

"And what are we going to do with him?" Xavier asked, waving the sidearm toward the guard. "He can't be out cold forever."

"I have an idea," Brianne said. "We've taken care of the phones. Emil needs to disable the computer so he can't email anybody when he wakes up. Now, somebody find me some heavy-duty tape. And I'll need everybody's loose change."

"Does anybody carry coins anymore?" Emil asked.

Brianne put a hand to his cheek. "Spoken like a true child of technology."

Brad handed her three pennies and a nickel. Xavier added a few more coins to the collection.

"This should be enough," Brianne said, adding her own. "Didn't you guys live in dorms in college? Ever heard of pennying someone in?"

Xavier and Emil looked as bewildered as Brad felt.

"I'll show you," she said cheerfully. Emil handed her some masking tape and she said, "Everybody out. Not you, kitty."

They all exited the room and Brianne shut the door, trapping the cat with her human. "Now," she said professorially. "If we force the door to bend out…" Emil and Brad shoved on the door. "…and shove the coins into the jamb like this—" She tapped the coins in until no more would fit. "It forces the door knob against the lock so it won't open. But if the person on the inside gives the door a few good hits, the coins usually come out, which is what the tape is for." Brianne taped the coins securely in place. She brushed her hands together and said, "Time to find Katya."

"Hey," Xavier said, "in the movies they always locate the prisoner by searching on a computer."

Emil said, "Yeah, and they have light sabers, too. If only." He let out another sneeze.

"It's worth a try," Brad said.

Emil scoffed. "Be my guest. But I think we'll have our best luck splitting up."

Brad agreed. "Why don't you two go into the sublevel, and Xavier and I will search up here, if we don't find something on a computer somewhere."

"Hey, wait a minute," Brianne said. "In the movies, something bad always happens when somebody says 'Let's split up'."

"I guess we're about to check your cinematic theory," Brad laughed nervously.

The door at the end of the hall opened. Brad desperately glanced around for a hiding place. A hallway split off to the left. "Quick!" he whispered.

The four stepped into the corridor, out of sight for now. They flattened themselves against the wall and watched as three guards and a guy in a lab coat passed by. Fortunately, they didn't look back. Once they were gone, Xavier said, "We need to set that fire ASAP."

"I guess that's up to me," Brad said, pulling a small incendiary device from his paper bag. "I'll go down that way. Brianne and Emil know what to do with the fire suppression system downstairs. As soon as the alarms kick in, we can search relatively undisturbed. Meet back here."

Brianne and Emil disappeared into the nearest stair well. Brad and Xavier sneaked down the corridor, farther into the depths of the warehouse. Most offices were darkened this time of night, but they began to hear the sounds of activity here and there: the tap of keyboards, the odd office conversation, phones ringing and computer workstation to station chats. They listened at every locked door, peeked down corridors and peered through interior windows. Brad constantly glanced behind himself and flinched at every sound.

At the end of the main hall, they found a double door marked *AUTHORIZED PERSONNEL ONLY*. Xavier gently pushed it open. Beyond lay complete darkness. Brad stepped in and turned on his cell flashlight.

It took a few moments for his eyes to adjust. At the edge of the darkness, on the far side of the room, two structures similar to ChronoCorp's pods glimmered in his phone's faint light. Beyond them spread another structure.

"What is that thing?" Brad asked.

"I'm pretty sure it's what the Ambassador called a chronometric coach. It seats twenty."

"*Twenty?* That was part of your phone call?"

Xavier nodded, but fell silent when two figures shadowed the windows in the double doors. The figures moved on.

"I hate to suggest it," Brad said, "but I guess we need to go upstairs."

"Time for our little fireworks display?"

Brad placed the device in an office at the side of the big room, set the timer for two minutes, and prayed Brianne and Emil had disabled the sprinkler system. Otherwise, it would be a very short search for Katya.

The alarms went off as planned. Brad and Xavier watched from an upstairs window as employees vacated the premises. They had to work fast, but the second floor offered no more satisfaction, and it was getting late. Reluctantly, they headed back to the ground floor. Brianne and Emil came through the stairwell door dressed as janitors.

"Nice fashion," Xavier said.

"You never gave us those ChronoCorp tee shirts you promised us," Emil said. "Besides, we wanted to fit in."

"I really don't think she's here," Brianne murmured.

"And we've got to get out of here before the fire department arrives," Xavier said.

Brad felt a profound sense of dejection. Where was she? He realized there was only one plausible answer. "The North Facility. That's where they've got to be holding her."

"Well, fine," Emil said. "But does anybody know the way?"

"Rex does," Brianne said.

"Yeah, but I'll bet he's too busy to answer his phone just now," Xavier guessed.

They stole quickly toward the door they had first entered. As they passed the room with the coins shoved into the jamb, they could hear the guard pounding on the inside and the plaintive mews of the feline.

70

Mila MIA

Mila van Dijk studied the Ambassador, a.k.a. Franklin Louis Bouchet. He was the picture of distraction, wringing his hands, his eyes darting from side to side. And he wasn't kidding.

"We must get rid of Katya, one way or another. And a few others, perhaps."

"I don't understand you at all," Mila said.

"There are large-scale things at stake here. Our entire plan is endangered because of ChronoCorp. More to the point, because of how unstable Xavier Stengel is. I offered to combine our forces and focus, but no dice. I'd rather negotiate with a baboon. But it's Katya Joshi and Todd Tanaka that have to be removed from the equation. Their expertise drives a lot of what ChronoCorp is all about."

"ChronoCorp's chief engineer, Todd, has died," Mila said. "Does that make you sad, Lou, or are you just too busy to feel something for a fellow researcher?"

"I will not dignify that with an answer."

"You just have."

"I'm not playing those old games with you, Mila. We need to take care of Katya. Along with that French technician."

"Emil?"

"Emil. Yes. I'll have Dresden see to it."

Mila looked at Bouchet. "You haven't changed. These people aren't individuals to you. They're disposable paper cups destined for the landfill."

Her ex-husband frowned at her. Mila wondered how he could view his fellow humans in this twisted way and still believe he could create a utopia for the world. It was a paradox—one that alarmed her beyond measure.

© The Author(s), under exclusive license to Springer Nature Switzerland AG 2021
M. Carroll, *Plato's Labyrinth*, Science and Fiction,
https://doi.org/10.1007/978-3-030-91709-8_70

* * *

Rex took a set of stairs into a third sublevel. The Imperium must enjoy deep basements.   Rex's phone buzzed again, another text from Bradley Glenn. He marveled that he had any reception here.

REX?

Rex was getting more and more nervous. The last thing he and Dresden needed was more distractions.

REALLY NOT A GOOD TIME

Brad's reply was instantaneous.

WE'RE ON OUR WAY, WITH OR WITHOUT YOUR HELP.

With a huff, Rex sent Brad the map, along with a warning.

JUST DON'T GET TOO CLOSE.

71

Spies and Bonfires

Katya was getting tired of the duct tape on her wrists, and she ached from the hard wooden chair (why couldn't it have been a fluffy recliner?). Her purse hung on the wall, well out of reach. The entire scene was right out of her favorite TV spy series, *Alias*. Too bad she wasn't as talented as her Sidney Bristow; with all that spy training she'd be out in a jiffy. But she tried to make herself content, assuming that Xavier and the others were negotiating her release. It had to be the Ambassador behind all this. She couldn't believe Mila would betray all of them. Her chest ached with the thought.

This was one of the places that needed to be disassembled, atom by atom, along with the Headquarters in Fort Collins. Luckily, her kidnappers had delivered her directly to where she needed to be. If she could free herself, she could explore and maybe, just maybe, figure out a way to bring the place down. But two cameras in the upper corners of the room were both active: she saw green lights glowing on their sides. If she tried to extricate herself, they'd send some Neanderthal posthaste. Still, she could loosen her bonds inconspicuously, and she had been working on the process for an hour.

A klaxon rang out, then another, as distant alarms blared all over the facility. Agent Bristow had gotten out of a similar situation by doing a reverse summersault over the back of the chair, landing on her feet, and bending over to release her tied hands. It looked easy.

Kat tried to swing her feet over her head, but she was bound to the chair seat. Her balance was all wrong. She leaned forward until the chair began to tip, and stood. Now she had leverage. She thrust herself backwards, but the chair swiveled to the side. She fell on her back, hard. She lay there a moment, seeing stars. She also saw the ceiling, and the cameras. They were no longer on.

M. Carroll, *Plato's Labyrinth*, Science and Fiction,
https://doi.org/10.1007/978-3-030-91709-8_71

Quickly, she rolled to the side. She slid the chair on the floor until her arms were close to her face. She bit through the tape on her wrist, a tear that would now easily rip. She tore the tape across both arms, then freed her legs, and stood. They would probably be coming for her soon, but there was one more move to make: she stepped back to the chair and stomped on one of the legs. It broke off at the seat. Grabbing the leg like a club, she lunged to the wall behind the door.

The alarms became louder as a guard stepped into the room. Kat swung the lumber at the base of his head. He crumpled. She glanced at his uniform. Whoever "Clive" was would not be happy with her once he woke up. She grabbed her purse and looked into the hallway. It was bedlam: people scampered in and out of stairwells and toward the exits. But she could smell no smoke. Perhaps she should go down to the basement and see what was really there.

* * *

With Rex's map finally in hand, Brad urged Emil and Brianne into his rental. If he drove, he hoped he would feel as if he was actively doing something to help Kat. Xavier, mercurial as ever, had gone off on his own to destinations unknown.

"So what do we do when we get up there?" Brianne asked. "The place must be guarded or monitored or something, full of security people who will want to see I.D. Did anyone remember their Primus Imperium badge? I mean, it's one thing to break into a semi-abandoned warehouse, but they make this place sound like a maximum security bank vault."

"We'll figure something out," Brad snapped, his eyes on the road ahead. But the truth was that he had no idea how they would proceed once they arrived. Katya Joshi could be anywhere, from some storage closet in the basement to a locked ward on an upper floor.

Every time he thought of her, held against her will, perhaps tied up or gagged, it drove him crazy. What if he never got to share a cannoli at their favorite restaurant again? He realized they hadn't even had time to establish a favorite restaurant of "theirs." What if that last cup of coffee they had together was, in fact, the last? He thought of her laughter, of the way she tipped her head to one side when she was teasing, of the way she glowed when she talked about her work. He gunned the gas. The tires squealed around the curves.

"Whoa there, cowboy," Emil said, grabbing the door. "Keep it between the lines."

As the rental rounded the next bend, an orange glow filled the roadway up ahead. Trees blocked their view of its source, but Brad knew they had nearly arrived. A driveway forked from the main road. He swung the car into it. He could see more light coming from up ahead, flickering through the trees.

"Hey, look at that," Brianne shouted. The gates were wide open. "Somebody up there likes us!"

As they approached, they could see a large parking lot. People milled around, hugged each other, or studied the wavering flames showing through the windows and spreading across the roof. Brad pulled off before they reached the lot, parking behind several scrubby trees. The trio paused at the edge of the lot, standing in the shadows. The sky had reached full darkness, providing a dramatic backdrop to the orange tongues of flame. Sparks danced against the firmament, mingling with the stars.

Suddenly, a loud crack issued from beneath the building. Deep rumblings seemed to push a wall of fire and smoke up through the windows. Fissures formed in the walls, gushing smoke and yellow trails of glowing meteors. Another loud bang, and then another, announced the next set of explosions, these more violent than the first. Smoke billowed, rising above the crumbling roof like a blooming orange flower. Brad heard a woman scream somewhere in the parking lot. He felt like doing the same. No one could survive such an inferno.

Several loud booms sent rumbles through Brad's chest.

"Hey guys, you made it." It was Rex, who had crossed the lot and found them under the trees.

Brad grabbed him by the shoulders. "Where's Kat?"

"Dresden went after her. Haven't seen either of them."

Brad watched in horror as the central hub of the building collapsed. The wings on either side followed suit in incandescent purple and orange whirlwinds of flame. Brad sensed a presence before he saw it. The large dark cowboy stood behind him. Next to him was a panting Katya. "That was close," she said.

Brad sprang to her and threw his arms around her. "I was afraid you…" His voice trailed off. He had no more air to speak.

Kat brushed a lock of hair from Brad's forehead. "I'm glad to see you, too. Thanks to Dresden, I—"

She turned, but Dresden had vanished.

"He does that," Rex shrugged.

The sirens of emergency vehicles were getting louder as they raced up the winding roadway. "Let's get you in the car," Brad said to Katya. "You need a ride, Rex? Climb in the back with Emil and Brianne."

"Cozy," Rex said, but he didn't sound like he was complaining.

Brad pulled onto the road and headed downhill toward the Fort. Kat laid her hand on his and said, "Brad, I'm worried that Xavier's planning something violent."

"More violent than arson?" Brad scoffed.

"Afraid so. But it looks like this North Facility has been permanently taken care of, and if we destroy the headquarters in town, along with their files, Xavier won't have to do anything. The pressure will be off and he won't resort to anything rash. Like picking people off."

"You mean killing them?"

She didn't answer, but her expression confirmed it.

"We loaded my worm into their network," Emil said. "Files should be scrubbed by morning."

"But we still have to get rid of the hardware," Kat said. "Their time pods and such."

"We burn down that warehouse and the entire industrial district of Fort Collins might go up in smoke," Brianne objected.

"Maybe we need something more measured," Katya offered. "As long as Emil's gotten rid of the software, it might be enough to destroy the time travel pods without worrying about the entire facility. But who's going to do it?"

72

Boom or Bust

Mila van Dijk was having a crisis of confidence.

Her ex-husband Bouchet had kept his dealing in illegal arms separate from their family finances. Their divorce had been "amicable," and after he had paid restitution—which was considerable—they split things down the middle, more or less, and Mila received a substantial fortune. As soon as the courtroom dust settled, she established a foundation for scientific endeavors in schools and universities, and in ventures in the professional sector (which overlapped the universities). Taxes were not the only reason for her generosity: she genuinely wanted to help the advancement of knowledge. She figured humanity needed all the help it could get.

As for Franklin Louis Bouchet, she really did love the man. *But what was he thinking?*

"Roman soldiers?" she bellowed at him. He faced her across his office, flinching as she continued. "You've got to be joking. I shouldn't be surprised; you always feel you can solve the world's problems with force."

He held up a finger. "Strategically controlled force."

"Oh, controlled, is it? Those armaments for Niger and Libya, and then the ones for Venezuela. How did that work out for you? But you didn't learn. Now it's a world police force manned by warriors from the Roman Empire. With you, it's all or nothing. You live at the farthest swing of the pendulum."

"Mila, my sweet, I know you're upset. But if you can just trust me for a while longer, you'll see the wisdom of it. War will disappear. The economy will be globally coordinated in ways far better than before. Peace will reign supreme. The *Pax Imperium*."

© The Author(s), under exclusive license to Springer Nature Switzerland AG 2021
M. Carroll, *Plato's Labyrinth*, Science and Fiction,
https://doi.org/10.1007/978-3-030-91709-8_72

"The Pax Imperium." Mila shook her head sneered at him. "Lou, this is too much. Nothing is worth this."

"On the contrary, for any great vision, many must make sacrifices."

"Not this one," Mila said coldly. "I quit our marriage; now I officially quit this, too."

"We need to work together, my dear. What would happen otherwise, if all those nasty secrets about Libya and Syria come out? That would be tragic to your foundation, wouldn't it?"

"Feel free to 'out' my terrible, dark dealings in antiquities. I can't stand by any longer. This has all gone too far."

She sped from the Imperium parking lot as quickly as her sports car would take her. She went east, the opposite direction that Louis would expect. Once she had reached a quiet little suburb, she pulled over and tried to call Katya. She assumed Katya would be well-supervised and phoneless and held out little hope of reaching her, unless Mila's guard had carried out her instructions. But Katya picked up on the first ring.

"Mila."

"Good to hear your voice. I had convinced the big guard, Clive, to let you go, but I didn't think he could do it so quickly."

"He didn't get the chance. I got out." Katya didn't elaborate on her brief interaction with the guard.

"I'm glad of that. I really am. I know it will be difficult to trust me right now, Katya, but you must believe me when I say I had no control over events at my home."

She said nothing.

"Katya, this is important: are you at ChronoCorp?"

"No." There was little cheer in her voice, but a good dose of suspicion.

"I called to warn you: don't go back there, and get Emil and Brianne and Xavier out of there, if they're at the offices. They're in danger."

Mila broke the connection before Kat could argue.

* * *

Rex, Katya, Emil, Brad and Brianne stood at the rest stop north of Fort Collins, forming a little semi-circle around Brad's rental car.

"If you want to shut the whole thing down," Rex said, "you'll need to get rid of more than their time pods. They've got a troop carrier the size of a Winnebago. But not to worry: while I was in the North Facility's basement, I did some shopping." He held up two compact packages. "These babies work wonders. I've used them before. Got a whole bag of them."

He handed one to Katya. "Explosives are not among my top hobbies," she said.

"These are easy. I can show you when the time comes."

"That time's gotta be soon," Brad said. "If the Imperium figures out who's behind the fire up the Poudre Canyon, and the little one we set at the warehouse, they'll be on our doorstep."

"Thanks to our visit to the North Facility, they've got me and Dresden and Katya on their CCTV feed, I'm sure," Rex said. "And that combination adds up to ChronoCorp."

"It's just a matter of time before they visit our place of employment," Katya said. She studied the small explosives Rex had handed her. "Rex, why can't you do it? You and Dresden?"

Rex shivered, glancing at the picnic tables and parked cars at the rest stop. "One act of arson is enough for me. The place is gone, razed to the ground. And that's not the only thing that's gone. Dresden's disappeared. Vamoosed. In some ways, I'm gonna miss that cowboy. The son of a bitch."

"What about Emil?" Brad asked. "Can he help out?"

"Nothing doing," Katya said. "I'm sending Brianne and Emil out of town for a while." She turned to them. "The two of you need to get lost for a couple months, for your own good. The Imperium is sending someone out to ChronoCorp for all of us, I'm pretty sure."

Rex could feel the hot breath of the Primus Imperium and, more specifically, the Ambassador, blowing down his neck.

Brad said, "Guess that leaves us." He met Katya's eyes. "I'll go with you."

"Brad, I can't ask you to do that. This isn't your battle."

"As soon as those thugs kidnapped you, it became my battle."

Katya grinned and looked at the floor. It was the first time Rex had seen her blush.

73

Bad News

The Ambassador's adjutant, Miss Graves, entered as if she was stepping before a firing squad. "Sir," she said, standing straight with military formality, hoping it would help.

The Ambassador looked up from his screen. "My dear, why so morose?"

"We've had some bad news, sir."

"Bad news is the nature of the world. We roll with it. Fill me in."

She flinched, not wanting to pop his bubble, then cleared her throat. "Sir, it's the North Facility. The armory has been destroyed. Actually, the whole thing has been destroyed. Explosives in the basement."

"The North Facility?" he whispered with a tone of incredulity. "Gone?"

"Yes sir. Burned to the ground. The gates opened just like they were supposed to, but it was too late. The fire department got there when the building was fully engulfed. It's already all over the internet."

The Ambassador hit a few keys and leaned in toward his screen. "Razed to the ground."

"The good news is that no one was lost. No casualties."

The Ambassador huffed, as if casualties didn't matter as much as the building and its armament stores did.

"Our people managed to corral all of our law enforcement trainees into a group before anyone was the wiser."

"Where are they now?" he asked, his head in his hands.

"They took them over to the rest stop north of town for a while, but they're on their way here now."

"That was quick thinking."

© The Author(s), under exclusive license to Springer Nature Switzerland AG 2021
M. Carroll, *Plato's Labyrinth*, Science and Fiction,
https://doi.org/10.1007/978-3-030-91709-8_73

"Yes sir. We have good people." Graves was putting off the second shoe-drop as long as she could, but the Ambassador saw through it.

"Was there something else, Ms. Graves?"

"I'm afraid so, sir." She paused for a beat. "Our computer techs say our system has been compromised. It appears that our master files have been corrupted."

"My God, this is a full frontal attack!" he thundered. "Can they be saved?"

"They say they are beyond repair, and some kind of virus has destroyed even the remote backup files. It went after them. Actively."

"Sophisticated." He let out a long breath. "How bad can things get?"

At that moment, the sound of an explosion roared through the halls. Then another. The Ambassador and Graves looked at each other and said in unison, "The Transference Theater!"

They scrambled into the hallway, meeting several people running away from the theater and several others moving toward it with fire extinguishers. Another blast set their ears ringing. As they approached the theater, the acrid smell of burning plastic and melting electronics filled the air. The room was a disaster. Both temporal capsules sat on the floor looking like broken eggs, and the troop transport was fully engulfed in fire. A man in civilian clothing was just escaping through the emergency exit, and a woman seemed to be making her way to a second exit across the room. But when she spotted the Ambassador, she turned and made a beeline toward him. Graves wondered if she was going to attack him. She tensed and placed her hand on her sidearm as the woman approached.

"You must be the Ambassador," she said. "Rex Berringer described you."

The Ambassador glanced beyond her to the destruction she had apparently wrought. "Who are you?" he demanded.

"I'm Katya Joshi," she said.

"Of ChronoCorp. Of course."

"It seems that your Imperium and my ChronoCorp are both out of the time travel business."

"You can't decide that."

Katya scoffed. "This from the man who is ferrying troops from another time into ours? I think you've lost some credibility, don't you?"

Another loud pop issued from the transport, and then a groan as it collapsed. Two guards aimed extinguishers at it while the building's fire suppression sprinklers began to rain over everyone, including the Ambassador.

Katya patted him on the chest as she passed him. "It's for the best." She ran out the door.

"I'll kill her," the Ambassador muttered. "Guards!"

A woman had just entered behind him. She was about the Ambassador's age with perfectly coiffed silvery hair and a classy riding outfit. She was clearly a civilian, and Graves started to block her path. The woman merely pushed her aside as if she was in charge. At that moment, Graves recognized the woman from various files and dossiers she had studied: the intruder was the former wife of the Ambassador, Mila van Dijk.

Mila put a hand on the Ambassador's shoulder, gently. "Louis dear, you don't want to start your new world with murder on your hands, do you?"

Graves had never seen the Ambassador like this. Perhaps it was the shock of the last few hours' events, but this woman seemed to have a unique influence over him. He looked around. "We've done so much work here. We built so much."

"I've seen you rebuild before." As she left, Mila called over her shoulder, "Consider it a challenge."

74

Uneasy Reunions

True to his word, Xavier picked up his twin the following afternoon. Even with the damage done to the North Facility and warehouse, he knew the facilities were beside the point: the prime actors were the ones he needed to take care of. He dropped Xavier Ω at the agency, where his doppelganger settled into a compact and headed for the foothills and Mila van Dijk's mansion. As Xavier watched the rental drive away, he couldn't shake the feeling that he had made a terrible mistake. But what was it? His mind was so muddled these days. It was difficult managing two of yourselves.

It wasn't until he returned to his office that he realized his error: Xavier Ω needed to be farther into the past for the van Dijk assassination to have the desired effect. Mila must not get the chance to fund ChronoCorp. How obvious! And yet it had somehow escaped his view. What else had he gotten wrong? He let out a moan, balled his fist, and pounded it through the drywall.

* * *

Xavier had done it. He wasn't sure he had it in him, but he had actually carried out another assassination. A murder.

The problem now was more immediate: he couldn't remember things. Important things. Like how he got back to this abandoned cabin, or what he had done with the gun. Or where he had gotten the weapon in the first place or where he was a week ago. Even more disturbing was the fact that he knew, beyond the shadow of a doubt, that he had a perfect alibi. What could that be?

He searched his memory. Why would he be so sure the authorities wouldn't come after him? He remembered something about San Francisco. About a

© The Author(s), under exclusive license to Springer Nature Switzerland AG 2021
M. Carroll, *Plato's Labyrinth*, Science and Fiction,
https://doi.org/10.1007/978-3-030-91709-8_74

meeting. Was he missing something? It was a conference. Yes, that was it. A conference featuring ancient civilizations. But the conference was days ago. During the murder. And he hadn't been there. He had been in Michigan, at the home of that brilliant researcher from France or India or somewhere. Still, the conference was the key. It was his out, his alibi. How could it be, if he hadn't been there?

He remembered bits and pieces, like a row of Egyptian sphinxes, and that big driveway. And Mila. That was it. He couldn't remember what Mila van Dijk looked like, but he knew what he was supposed to do. He had shot the little rich man out in his yard—not in the head as the other Xavier had requested, but in the back. Probably through the heart. It would do. He had killed Mila van Dijk.

What kind of name was Mila? Shouldn't it be Milo or something?

In the back of his mind, just at the edge of awareness, he recalled the other Xavier, the one that called him Xavier Ω. He glanced through the dusty window into the gravel clearing. His old car came bouncing up the dirt road, tossing a rooster tail of gravel behind it.

Who was driving his car? And how did they get the keys?

It pulled to a stop in a cloud of dust. A person stepped out. There was something familiar about his demeanor, how he held himself. And that shirt he wore. The guy looked up toward the cabin. It was as if Xavier was looking into a mirror.

It was then that he remembered that he was the fall guy. He had been doing someone else's work, and it had been dirty. He dropped to the floor and peered through a gap between wallboards. His doppelganger walked up toward the cabin, glancing from side to side as if looking for a sniper in the trees. From his jacket, the man pulled out a sidearm.

Xavier turned on his knees, trying desperately to make no sound, and crawled toward the back. But isn't that what anyone would do? Head for the back door?

He heard a rustling behind the cabin. The twin would break through that back door any second. He had only one chance, and he took it. He dashed across the cabin, dove through the front door, and prayed that the other Xavier was still beyond the back side of the building. He heard the door crash from behind as the intruder broke into the shack. He leaped across the road and lunged through the open car door. The car was idling. He dropped it into reverse, gunned it, and raced down the hill backwards until he found a turnout. The tires squealed as he swung the car around. In moments, the cabin had vanished from the rearview mirror. Just like his past.

* * *

Xavier Ω drove blindly for half an hour, until he realized he had ended up on Xavier's street. His street. He pulled into the driveway. The other Xavier, the one from this universe, would be stranded up at the cabin, giving Xavier Ω a head start. He realized something else: the ring hanging from the ignition had a bouquet of jangling keys, one of which was probably the key to Xavier's door.

He shut off the engine, pulled the keys and walked across the front lawn. Everything felt familiar, as if he had made this walk a million times before. He held up the keys and knew which one fit.

"Xavier?" came a voice from next door. It was the 80-year-old grandma who had planted flowers well into his property. He didn't really mind. She peered at him from beneath a sprawling sun hat.

"Hello, Esmeralda. How are the hydrangeas?" Hydrangeas? Esmeralda? Memories were coming back, solidifying like ice on a lake, becoming thicker and more stable, a place he could walk upon with confidence.

"Still infested, but I've got some new stuff that may do the trick. I wanted to tell you something." She came to the edge of her yard and beckoned Xavier Ω over. He approached.

"What's up?"

"Your place was crawling with cops a couple hours ago," she whispered confidentially.

Xavier forced a laugh. "Case of mistaken identity. They thought I was someone else."

"That's a relief. I knew it was something like that."

"Anyway, good luck with your bugs. Hope you can get rid of them."

She waved her spade at him and returned to battle the insect world.

Xavier Ω opened the door and stepped inside. He closed and locked the front door, and then stood with his back against it, scanning the room. There was the couch, the new large screen on the wall, the gas fireplace and chairs. At left stood the rolltop desk he had been given by his grandparents.

Grandparents. He thought about the grandfather paradox. *What would happen if I traveled back in time and killed my grandfather? Then there would be no me, and no one to kill my grandfather.* And yet, Xavier had somehow short-circuited those strange paradoxes, had broken through time to find himself. It was all so baffling. One thing was certain: for the time being, he was stranded in this parallel world. If only he had some resources, some money.

But wait: he had access to all the money he needed! He opened the rolltop and slid a small drawer open. Inside was the bankbook he knew would be there. And the passport. In the end, he was essentially taking it from himself, wasn't he?

Now to another problem. Where could he travel that didn't require a passport? Even Canada and Mexico were becoming twitchy about such things. Bob would know. Robert Shiloh, his old school buddy, had played running back to Xavier's quarterback. Back then, they were thick as thieves. These days, Bob had a deep sea fishing business. People chartered his boat all over the Caribbean, the Atlantic coast, and the Gulf of Mexico.

Xavier Ω had lost his cell phone somewhere along the way but there was still a landline in the office. He looked up Bob's number, dialed, and made some small talk, hoping he didn't say anything that diverged from this reality that Bob inhabited. Finally, Xavier popped the question.

After a pause on the other end, Bob said, "I think I can arrange that for an old college football chum. Do you still throw the pigskin around?"

"I'm afraid those days are over. Arthritis in my shoulder."

"I'm past my football days, too," Bob lamented. "I've got a torn ACL that keeps getting my attention. Hey, you're too young for arthritis. Like I said, with your height, you should have gone in for basketball. Not that you weren't a fine QB. So how does a Caribbean island sound?"

75

Off-Season Christmas

Sergeant David Bluedeer sat across the wooden table from Mila van Dijk. His associate, Jimmy Fjell, tapped at a computer screen. The room was dimly lit. Van Dijk had a glass of water next to her, adding another circular scar to the battered tabletop.

"Now," Bluedeer encouraged in a quiet voice, "this was taken by the traffic cam mounted on the light just down the street from your house. Just take your time. We can go over it as many times as you want. Jimmy?"

Fjell tapped his screen. The large monitor in front of Mila sprang to life. A fuzzy image of her front drive wavered with sculpted bush animals along the street. Just visible on the left edge of the screen stood one of the black sphinxes leading up her front walk.

"Wait for it." Bluedeer intoned. "There. Here comes the gardener."

"Yes," Mila said fondly. "Barnes."

Bluedeer nodded toward the monitor. "He begins trimming the hedge, and…"

A car pulled to the curb. The driver stepped out, swinging a gun up with both hands, braced against the roof, and shot. Mila let out a moan.

"Sorry, Ms. Van Dijk, but your help is critical," Bluedeer said as the video showed the shooter getting back into the car. He sped away. Barnes lay on the grass, not moving.

"Mr. Barnes was a very lucky man," Jimmy said.

"Yes," Mila agreed. "The doctor said that had the bullet gone in just an inch below, it would have hit his heart; just missed some important vein or artery above it."

M. Carroll, *Plato's Labyrinth*, Science and Fiction, https://doi.org/10.1007/978-3-030-91709-8_75

"Yes, yes," Bluedeer said, keeping the impatience from his voice. "But the shooter?"

Mila shook her head. "It's a difficult angle."

"Would you like to see it again? He turns at the last, just before he slides into the vehicle."

"Sure," she said. Bluedeer noted the lack of enthusiasm. Perhaps the scene of her own employee's attack was too much for her to witness, but he suspected it was something else.

They ran the CCTV again, and Bluedeer watched Mila carefully. The light from the screen flickered across her face, painting it with cadaverous tones. When the footage reached its moment of clarity, her expression changed subtly. Could Bluedeer see recognition there?

"So?" he asked.

Mila leaned back in her chair and rubbed her eyes. "Sorry, Sergeant. It *could be* Xavier Stengel, but I just can't say for sure."

Jimmy brought the file up to the moment the culprit turned toward the camera and hit pause. Bluedeer said, "And you're sure."

"Yes, I am. I wish I could be more help." Mila didn't meet his eyes when she said it.

"Shall I run it again?" Jimmy offered.

"I don't see the need," Mila said.

"Nor do I," the Sergeant said. Jimmy glanced at him with a puzzled look, but Bluedeer was already on his way around the table to pull Mila's chair out.

"I can't thank you enough for coming in," Bluedeer said.

"Yes, thanks so much," Jimmy added.

Bluedeer escorted her to the front exit and returned to the room, where Jimmy pounced on him.

"Geez, Sarge, you let her off easy. 'Don't see the need' to look again? Why not make her sweat a little."

"First of all, people like Mila van Dijk don't 'sweat.' She's spent all her adulthood learning how to be smooth and calm and enigmatic. You can tell; she's very polished in her manner. Second, there was no need because she did, in fact, recognize the person. At least that's my gut. Whoever it is, she's not going to throw them under any bus."

"Protecting someone?" Jimmy asked.

"Maybe. The first time through the footage I saw some well-masked surprise, and the second time there was clear recognition. But she's good. I think we've gotten all we can here. But we've got enough on Stengel as it is. With his tie-in to the attempted murder in Michigan, the DA can lock him up until Hell freezes over."

"Sarge, you've seen the guy. He won't be in any prison. An insanity plea comes to mind."

"We'll just let the cogs of the justice system turn."

"You're the boss."

"Yes," Bluedeer said. "Yes, I am."

* * *

Louis Bouchet had very little time to consider Mila's challenge to rebuild his vision; he was soon given a new home in the "Supermax" maximum security prison in Florence, Colorado. His possession of the military-grade weaponry found in what was left of the basement was a serious affair and a breach of his parole. Combined with several counts of attempted murder—the product of plea bargains by a few former employees—his stay would be a long one. His Primus Imperium was effectively in the past.

The dust and smoke and detritus of the North Facility's demise seemed to settle quickly, and Mila van Dijk invited Katya and Brad to her home, this time for a luncheon. She explained that she wanted to have the get-together before Brad returned to New York, and she included Rex Berringer in her guest list because of his good work with Katya and ChronoCorp. The gesture was an olive branch, but Katya couldn't bring herself to trust Mila. Not just yet.

Mila stirred sugar into her tea with surgical precision. She tapped her spoon on the rim of the china cup and looked up. Between her and her guests lay an impressive spread of sliced cheeses and meats, platters of fruits, and tureens of vegetables. An enormous basket overflowed with rolls and loaves of various breads. "I was very sorry to hear how things worked out with Xavier," she said. "I felt he was a good man."

"He is a good man, yes," Katya said.

"Still, it all ended tragically for him, locked up like that."

"He's not exactly locked up," Katya said. "The center is really beautiful: gardens, an Olympic-size swimming pool, rec center. It's not some medieval Bedlam."

Mila looked down, shaking her head. "I'm not proud of what I did. You must understand: my former husband was threatening to ruin me."

"Xavier's not crazy, you know," Katya said gently. "When you have a broken leg you have to rest it. A broken mind needs rest, too."

"The Caesar salad is excellent," Brad said incongruously, speaking around a mouthful of greens. Katya fought back a smile and looked at Mila, who began to grin. Rex watched Brad, stifling a chuckle. In moments, laughter broke out around the table.

A domestic came into the dining room and poured a fine wine for them. Katya hoped it would help relax the mood even further.

"I'm sorry, dear," Mila said to Katya. Katya didn't like the informal way Mila still addressed her. ChronoCorp's benefactor hadn't earned her trust back that easily. "I didn't mean to imply that Xavier is crazy. But the world is crazy, certainly. Just the other day, my gardener was shot by some deranged stranger in my driveway. Why would someone do that? The police are looking at some CCTV recordings as we speak." She stood and walked to a cabinet.

Katya and Brad looked at each other, mouths open, as Mila continued. "But here, I want to make up a few things to all of you, to pay you back for the wonderful work you've done and the missteps I took. I know I almost made a real mess of things, and I'm sorry." She stepped over to a glassed breakfront against the wall. "This is a sort of cabinet of curiosities—precursors of modern museums. People would display their collections of objects from nature, like bones and shells and sculls, or artifacts or religious relics."

"Sounds like it would be hard to dust," Rex said.

As Mila opened the cabinet, Katya was reminded of a scene from the Wizard of Oz. *This brain is for you, Scarecrow. And this heart is for you, Tin Man. Have some courage, Lion.* She felt like Dorothy, who got nothing from the Wizard's treasure chest but a missed balloon flight.

"I've pulled a few select items from my collection," Mila said. "For Dr. Glenn and Mr. Berringer, I want the two of you to have these genuine Minoan clay beads. These are from about the period that your visits took place. If you hold them up to the light just right, you can see remnants of paint clinging to the sculpted pattern."

Brad studied his carefully and said, "I see a blue line, and patches of white. Very beautiful."

"Yeah," Rex said, struggling to see the same in his. "Thanks a lot." He stuffed the precious artifact in his pocket.

"Katya, I'd like you to keep this for Xavier. Save it for when he gets out. It's a stoneware cup, and you can see an inscription in Linear A on the inside lip."

The cup was exquisite, yellowish in color, and slightly translucent.

"And this," Mila said, "is for you." Mila handed Katya a shadow box. Beneath the glass lay a white figure of a priestess or possibly a bull jumper.

The little sculpture took her breath away for a moment. When she had recovered, she asked, "Ivory?"

Mila nodded. "Elephant. The bodice is painted on, although some figurines seem to have had some kind of leather or gold leaf for clothing. Why am I telling you this? You're the expert."

"Not in this area. Mila, these things are beautiful, and we will cherish them." She glanced at the two men.

"Yes, we will," Brad said on cue. "This is going in my collection of most prized objects from digs. It qualifies, I'm sure."

"It does indeed," Mila said. "There is just one more thing, Katya." She pulled a frame from her cabinet. The small square was no more than four inches across, but what it held was remarkable. Fragments of plaster from a wall painting had been lovingly reassembled. The picture they formed was a portrait of a woman with the typical black, curled hair worn by the Minoans, the low bodice exposing the breasts, and the sacral knot at the back of her head. Her lips stood out in ruby red. She reminded Katya of le Parisienne, or of the Priestess Kitane.

Mila put a hand on Katya's. "I know he is gone, but this is for Todd. I want you to have it."

Katya looked at the ancient painting and nodded silently. She knew if she looked at anyone just now, the prickling behind her eyes would become a flood. "I appreciate it, Mila," Katya said quietly.

* * *

"Fine with me if someone wants to soothe their guilt by handing out priceless artifacts," Rex said as Brad's car took the last sweeping turn out of the foothills and into Fort Collins proper. Rex sat in the back, with Katya riding shotgun.

"Oh, it was more than guilt, I'm sure," Brad said.

"I wonder if Franklin Louis Bouchet is feeling guilty at all," Rex mused. "He owes a lot to a lot of people," he said, crunching on a carrot he had taken from the dinner table.

As they entered the Fort, suburban homes passed by. Apartments for student housing rose up to the left, while a park and stores lined the street at right. Traffic came to a halt at a light, and Rex watched as a tour bus of some kind pulled up across the street. A large group of men stepped out wearing mismatched clothing, rumpled shirts and shorts, hats askew and strange combinations of socks. Rex could swear he recognized several of the big men, but he couldn't place them. Surprisingly, they were all speaking Latin. Perhaps it wasn't such a dead language after all. As they shuffled across the street to a Marine recruiting center, Brad drove on.

Rex looked toward the front, between the seats. Katya had reached over and was holding Brad's hand. Rex thanked the gods of the roadways that Brad hadn't rented a standard.

He leaned forward. "Hey guys, why don't you two drop me off at my car and then get a room somewhere?"

Katya twisted around to look at him. "And what about you, Rex? What's on your personal horizon?"

"Me? I'm off on another adventure, as far from the Ambassador as I can get. I'll send a postcard."

76

Conversations in the Dark

"I like your place," Brad said, taking another sip of wine. He sat on Katya's couch, the lights low, his thigh against hers.

"When do you leave this time?"

"Eight a.m. It's getting so that I can fly to New York with my eyes closed."

"Want a ride to the airport?"

"Already booked one, but thanks."

She leaned her head on his shoulder. "You've been quite gallant lately. Saving the damsel in distress and all."

"You don't strike me as the damsel in distress type. You can take care of yourself. I've seen you in action. And heard stories."

She straightened and looked him in the eye. "From Dad?"

"Ajit told only good stories. Don't worry."

She put her wineglass on the little table in front of the couch and rested her head on Brad's shoulder again. "You know, at one time I dreamed of going back and changing history. Telling ancient physicians about germs, and showing the doctors of the 19th century about DNA's double helix. Putting medicine far enough ahead that my mother survived her cancer."

"It's a beautiful vision," Brad said.

Katya shook her head. "The small changes we make just go away, and the big ones cascade into disaster. Whoever is up there pulling strings seems to have a better map than we do."

"Now you're sounding more like Todd." Brad smiled. "You're not thinking of becoming the next friar of ChronoCorp, are you?"

M. Carroll, *Plato's Labyrinth*, Science and Fiction,
https://doi.org/10.1007/978-3-030-91709-8_76

"A lot of Todd's ideas are beginning to make sense to me these days," she admitted.

"I'll buy you a cowl. I'm sure your father is proud of the work you've done here."

"He did say—"

Brad interrupted her with a kiss, long and gentle. She kissed him back.

77

Moving Experiences

Katya's life spun in an uproar. So much uncertainty surrounded her since the fall of ChronoCorp and the Imperium. Brad was back in New York. She wondered when he could carve out time to come back. She might have more room in her own schedule, now that her place of employment was essentially shut down. Maybe she should think about going back there for a visit.

The boss was gone, they had lost Todd, and no one wanted to reboot ChronoCorp in its current state. Xavier had announced with great fanfare that time travel research would be coming to an end for the foreseeable future. Recent events had seen to that. When she had visited him, he had been so drugged that he scarcely acknowledged her. The attending psychiatrist explained that Xavier had been violent, screaming about twins and other universes. The only semi-cogent thought Xavier had communicated to her during her visit was, "You can't get good help these days, even if it's yourself."

Like most of her fellow employees, Katya felt rudderless. She began to search the universities for teaching positions, but her passion was research. Where could she go? Berkeley Quantum? MIT? Fermilab? Argonne? Perhaps there was something closer to New York City…

Most of her fellow employees had moved out, and the lease on the ChronoCorp site was up in a month. She stared at the cardboard box she had put on her desk. Carefully, she began to place her personal items into it: a photo of her father, a bonsai tree with more brown leaves than green ones, a coffee mug that admonished, "Think like a proton: stay positive!"

She closed her door gently. Finally, Katya let herself release what was inside. She wept with sadness for the aching loss of Todd. She wept for Xavier's pain, and the struggle he faced on his road back to sanity. She wept for those left

M. Carroll, *Plato's Labyrinth*, Science and Fiction,
https://doi.org/10.1007/978-3-030-91709-8_77

behind, for the Minoan princess in the dark room, tending the musical movements of her clockwork mechanism until the last moments of Thera. She shed her tears for her father going to bed alone at night, and for her mother. And she wept with a deep contentment—was it joy?—for what she shared now with Brad. Finally, she understood why her Dad liked the guy so much. She let it all rise to the surface and exhaled it into the universe that was, for now, a safe place.

She heard a knock on the door. Wiping her eyes, she opened it to see Emil and Brianne standing in the open doorway.

"You're back!" Katya barked.

"We figured it was safe enough," Brianne said, "what with the Imperium all dismembered and the Ambassador locked up."

"So what are your plans?" Emil said, staring at the cardboard box.

"I've been looking into teaching."

"A noble cause," he said.

A voice from the hallway called out, "Emil, do these terminals go, too?"

"We didn't ask permission for those, so not yet," he said to someone down the corridor. A guy in a WE-MOOV-U uniform passed pushing a cart full of electronics. Katya hoped his company was better at moving than they were at spelling.

Katya frowned. "So what's with all the packing?"

Emil looked at Brianne. She looked at him. Emil said, "That's what we came to chat about."

His resolve seemed to ebb, so Brianne took up the torch. "Katya, we're transferring our lab to the University of Colorado in Boulder. Not everything, but the stuff pertaining to Emil's research."

Emil became animated. "See, we'll do more sedate work than we did here—no running off to ancient places—although we'll still be doing cutting edge stuff. And we were thinking the perfect mix would be Brianne and me and you."

"CU Boulder," Katya murmured. "That sounds pretty appealing, actually."

"We were hoping you'd say that," Brianne said, raising an eyebrow. "We're setting up our lab now. It's a bit smaller than we're used to, but we can still do good work. The three of us should go down and have a tour, and then lunch at the Pearl Street Mall."

Katya grinned. "I have a suspicion we'll be seeing a lot of Pearl Street in the future."

Brianne and Emil left to supervise the movers. Katya closed and locked her door. She sat at her desk, scanning the room, looking at the framed photos and cartoons, the various awards, and the little post-it notes nagging her from every corner of the room. Most of them meant nothing now. She tried to tell herself that Bradley meant nothing, too.

* * *

Bradley Glenn sat at his computer, staring at the job announcement. His mind was eighteen hundred miles to the west. Imagine: a brand new museum in Longmont, Colorado, dedicated to dinosaurs. It would be a tabla rasa, a blank slate to transport patrons into the Mesozoic. And they needed a curator. It could in a sense be his own time machine. He fit the bill, and had the right experience and training, but would it look too forward to Katya?

The phone rang. Katya's voice came through. "How's my favorite time traveler?"

Brad studied his computer screen again, thinking of the possibilities. He said, "I was actually thinking about you. There's a job possibility in Longmont that has me intrigued."

"Do tell."

"I'd curate a dinosaur museum there. Build it from the ground up. Would that be too weird? Me moving out there? I mean I'm really not stalking you or anything."

"I'm disappointed that you're not. Actually, I'm moving to Boulder to work at CU. It's less than a half hour drive from Longmont."

"It would be even shorter if I lived south of town."

"And if I lived north of Boulder, who knows? Maybe you should fly out again. We can discuss prospects."

He booked a ticket as soon as they hung up.

78

South London, 1880

Waterhouse Hawkins entered his studio in West Brompton. He was tired. His wife was ill, perhaps dying, and he needed some distraction; he had been at her bedside for days. Now, he struggled to remember something. So many of the details of his trips overseas seemed to be fading from his mind. His lovely statues in Central Park, his lectures, his murals at Princeton, and that mysterious woman he had met one night, a decade or more ago. He fought to recollect the particulars. He could remember a blue light. He could recall a cracked valise. But what had she wanted to show him?

He pulled the pince nez glasses from his pocket—the latest John Ourrin fashion in New York and London—and placed them on the bridge of his nose. He wanted to draw a new kind of dinosaur—something he dreamed about—but he just couldn't remember the details...

* * *

Bradley Glenn packed up the things in his desk at the museum and wished his colleagues goodbye. It was a bittersweet moment, but he was excited about his life ahead in Colorado, and his life ahead with his new fiancé, Katya Joshi. He drove back to his condo and carefully wrapped the items, placing them in boxes to ship. They would come in handy at the new museum.

Katya seemed perfectly at home talking about extra dimensions and microscopic vibrating strings, but the only vibrating strings he was comfortable with were those on violins and cellos. Time's currents and repetitions, its *echoes*—as Kat put it—the paradoxes, and the flow of history; he wondered how these ideas would affect their relationship. ChronoCorp's time travel had

© The Author(s), under exclusive license to Springer Nature Switzerland AG 2021
M. Carroll, *Plato's Labyrinth*, Science and Fiction,
https://doi.org/10.1007/978-3-030-91709-8_78

brought them together. Ancient Thera and Kat's kidnapping had deepened their relationship. What lay ahead was anyone's guess; life was an adventure.

Something caught in the folds of his coat. He reached into the pocket and pulled out a pair of glasses. Pince nez. They were Oakleys (the hottest in New York fashion, he had heard). "How did those get there?" he mumbled. He put them on, a reflex. They fit. Surprisingly, he could see better through them. What was it he had been wondering about? Something about his coat pocket. No matter.

79

Sand and Sea

Rex knew just where he wanted to settle down. The Bahamas had a nice sound to it, but it was really an archipelago of zillions of islands. Where to go? Fortunately, he had taken part in a discussion of this very subject with his friend with the pilot's license. With the funds that Xavier had paid him and the cash he'd put aside, he topped off his buddy's fuel tank and they took wing, eastbound out of Florida.

Rex had shopped for the best stays and the best beaches. He had finally decided on Guana Cay Beach, a fairly secluded two-mile stretch of powdery white sand on the island of Abacos. There were condos to rent and a "stellar" beach bar, as the travel brochures put it. Nipper's was indeed stellar, so Rex rented a condo nearby. He would look into more permanent real estate soon enough.

He stretched out on the sand like a marooned, badly sunburned walrus. He gazed over his toes toward the glistening blue-green water, listening to the distant steel band at Nipper's and the scream of the Frigate Birds overhead. It was the perfect place to reflect on events of the last months, something he seldom did.

His most recent and probably last job had to be the strangest one he had ever taken. He looked at the curling waves and remembered Xavier talking about time having currents, about how it could take you back to the same place over and over. Not coincidence. *Synchronicity.*

He took a sip of his drink. He had forgotten the name of the concoction, but he liked the slice of pineapple, the cherries and the two umbrellas sticking out. Closing his eyes, he took in a deep draft of the salt air, letting it mingle

M. Carroll, *Plato's Labyrinth*, Science and Fiction,
https://doi.org/10.1007/978-3-030-91709-8_79

with the tang of the alcohol. Yes, it was the perfect place; no one from that past life would ever find him.

"Rex?"

The voice sounded familiar. Rex yanked off his sunglasses, furious for the intrusion. But anger turned to shock. "Dresden?" he cried. "What are you doing here?"

The man had swapped his cowboy hat for an island-style bleached canvas one. It gleamed in the sunlight. He carried a chihuahua in one hand and a piña colada in the other. "It just seemed like the place to come." Then Dresden did something Rex had never seen: he laughed.

"Hey Dresden!" came another voice from the direction of Nipper's. "Wait up!"

A figure limped up the beach, tiptoeing on the hot sand. As he approached, Rex realized that it was Xavier.

"Hey guys," Xavier said. His eyes were clear, his speech strong, his manner sure.

"You look like a new man," Rex said.

"If it wasn't such a cliché, I'd say you look like you've just seen a ghost," Xavier said. He turned to Dresden. "And who is this?"

"Rex," Dresden said, "You must forgive Xavier, here. This is not quite who you think it is. This is Xavier, but it's not. It'll take some explaining."

"A few too many visits to Nipper's?" Rex offered.

"I've actually sworn off drinking for the foreseeable future,' Xavier replied.

"Haven't we all," Rex said, rubbing the back of his sandy neck.

"I'm afraid the Xavier you know hasn't fared so well," Dresden told Rex. "He was caught on CCTV shooting the gardener in front of Mila van Dijk's house. Last I heard, he's currently in a psychiatric facility."

"Well, guys," Xavier Ω said, "I've come into some money, so the next round's on me. Ginger ale for myself. May I take your orders?"

* * *

She was invisible in the darkness of her bedroom, but he sensed her there, the elegance of her aroma, the electricity of her presence.

"Kat," he said.

"Yeah?" she spoke into the darkness. He rested his hand on the curve of her hip.

"Just making sure you were there."

"Always."

They lay there, not needing to say anything. Two time-travelers who had ended up in this one place, this one moment. Brad leaned into her, and she into him. He kissed her on the neck below her jawline. He was self-conscious about his sandpaper cheek as it slid up against her neck. Her lips drifted to his and he let them. Soft, moist, warm. For just a moment, he thought of the bioimager.

"It's a little more relaxing without falling statues and exploding mountains," he said, very low, very quiet.

"Mmm," she murmured. "It's sad how it all ended."

"Ended?"

"Our temporal research. Kaput. Finis."

Brad shook his head with a soft smile. "No, it's not gone for good. Your work will be like those statues of Waterhouse Hawkins. It's buried for now, but one day it will surface again."

"Maybe."

He grazed her cheek with his finger, kissed her gently on the lips, and reveled in the fact that in all that universe out there, the spinning galaxies and the cosmos-warping black holes, the Colorado thunderstorms and the bizarre beasts of the Mesozoic—the ones of Hawkins and the ones of reality—time and space had brought the two of them together. Whether time travel was a good idea or not, Brad knew that his running into Katya was a very good thing, indeed.

Part II

The Science Behind the Story

80

The Science Behind the Story

Quantum Physics and Relativity

Big Picture: The Nature of the Universe and String Theory

Sir Isaac Newton (1643–1727) was a brilliant mathematician, physicist, and inventor. His equations became the guiding description for the laws of motion in our universe. Newton's math perfectly describes the universe we experience: height, width, depth, and time. His equations can be used to define how objects move (for example, the accelerating speed of a falling apple or the result of two cars smashing into each other). But Newton's numbers could not adequately portray the universe of the very small: atoms and subatomic particles. That math came from later scientists, including Albert Einstein and Werner Heisenberg. Their mathematics defined quantum mechanics. Einstein's general relativity saw the interaction of subatomic particles as occurring in continuous, smoothly interacting events determined by local conditions. But in quantum mechanics, these tiny events happened in abrupt jumps whose outcomes could only be predicted in probabilities rather than in definite results.

Around the beginning of the twentieth century, it became clear that Newtonian mechanics was only approximately correct, and Einstein's theory of special relativity worked better for objects moving at high speeds. Then, in the first few decades of the twentieth century, physicists found that the equations of motion also broke down for very small objects like subatomic particles, so they developed new set of mathematical laws: quantum mechanics. Physicists began with a straight-forward version that was non-relativistic (it

© The Author(s), under exclusive license to Springer Nature Switzerland AG 2021
M. Carroll, *Plato's Labyrinth*, Science and Fiction,
https://doi.org/10.1007/978-3-030-91709-8_80

operated in situations occurring below the speed of light). These calculations worked well for atomic systems. But their problems were not over. By the 1930s, it was clear that they also needed a relativistic version. Over the course of several decades, they developed quantum field theory, which correctly describes all the observed elementary particles at all the energies we can achieve in the laboratory. It also describes three of the four forces we see in nature—electromagnetism, weak nuclear forces (the ones that cause radioactive decay), and strong nuclear forces (forces that hold the nuclei of atoms together). Frustratingly, quantum field theories of point particles simply didn't work to describe gravitational forces.

Then came the string theories. "String theories hold out the promise of giving us a quantum theory of gravity," comments physicist Randy Ingermanson, "but the unexpected side effect is that a realistic string theory requires ten dimensions." Since we only see four large-scale dimensions, string theories would require that a further six dimensions curl up into very small structures that we can't easily observe.

To put it more simply in this framework, in order for both Einstein's and Newton's math to work together, the universe must have at least ten dimensions, as we will see below.

Thinking Threads

The four-dimensional universe that Newton knew can be described as a vast plane. Physicists refer to this plane as the "brane" (short for membrane). We can envision this brane as a grid of lines, with each square in that grid made up of smaller and smaller lines. As we get to the smallest of these lines, we see that each is made up of squiggles or loops that move above and below the brane. These filaments—or strings—represent other dimensions.

The idea of string theory is not new; scientists have tried to define its details mathematically for years. Initially, they failed, but a new version of the concept has brought new insights and solutions to the earlier vision. It is called superstring theory. Its story goes like this: molecules are made of atoms, and those atoms, in turn, consist of subatomic particles like the neutrons and protons that make up a nucleus, and are orbited by electrons. Even smaller are the quarks that make up the neutrons and protons. But superstring theory goes even deeper, suggesting that those quarks are made of dancing filaments. For those filaments, or strings, to exist, there must be other dimensions. The mathematics of superstring theory, like those of the earlier string theory, confirm that the universe must have at least ten dimensions of space and one of time.

Superstrings can be "knitted" in five ways, called Type I, Type IIA, Type IIB, SO [32] Heterotic, and E8 x E8 Heterotic. Their differences are not important to our discussion. What is important to understand is that all require the existence of six extra, unseen space dimensions. In terms of their mathematics, each of these versions tell the same quantum story from a different perspective.

If superstring theory is more than an abstract concept—if it can, in fact, describe the universe we live in—then it must be testable. Scientists are nearing ways of carrying out such investigations at several research centers. One center is the Large Hadron Collider near Geneva, Switzerland. The 27-kilometer (17-mile) ring is the world's largest particle accelerator. There, researchers can measure the energy of two colliding particles. If some of the resulting energy flies off into extra dimensions, the change in energy in our four dimensions can be detected.

Not all scientists embrace string theory. Without knowing the geometries of the extra dimensions in which these strings might operate, the theory's details cannot be verified, they object, and predictions cannot be tested.

The Many Worlds Concept

The existence of superstrings naturally leads us to the idea of multiple dimensions beyond the four that we experience. But what about multiple universes? Physicist Hugh Everett first proposed the Many Worlds Interpretation of quantum physics,[1] now more popularly called the parallel universe or multiverse theory, in 1957. When Everett developed it, he was trying to explain one of the most bizarre concepts of quantum physics—what is known as the collapse of the wave function. This aspect of quantum physics implies that whenever observations are carried out, all the possibilities collapse into one. Many potential outcomes exist, but as soon as an observer makes a measurement, they all collapse into one solution.

Physicist Randy Ingermanson explains, "you have a set of possible measurement values that could potentially come out of your experiment, and one of them gets selected at random when you actually make the measurement. You can prepare the experiment the same way many times, and you get a range of different measured values. This is extremely disturbing."

Everett argued that rather than collapsing into one possibility, all the possibilities play out in different, parallel universes. All the other outcomes still

---

[1] Everett, H., 1957, 'Relative State Formulation of Quantum Mechanics', *Review of Modern Physics*, 29: 454–462

exist in different analogous universes. This line of reasoning has led to the idea that some actions in a changed timeline will "imprint," while others will simply disappear, as we saw with our story's Hawkins drawing of the Iguanodon.

Todd says that, "There may be some evidence hinting at a few places where other universes bump into ours. If you sent someone a short enough throw, say, a few minutes into the future, their universe and yours might coincide." Later, a crazed Xavier makes use of merging universes, meeting his own doppelganger from another dimension. But is there any evidence that other universes actually exist? If so, how would we know?

One way of detecting a universe outside of our own is to search for places in our universe that are "bruised" or disrupted from the outside. The hunt for such disturbed cosmic spaces has been on since we began mapping the echo of the Big Bang. Throughout the universe, background radiation simmers at just three degrees above absolute zero. This cosmic background radiation is the fingerprint—or echo—of the Big Bang. It shows us that the universe is very smooth.

The explanation of the consistency throughout the universe is the result of something called inflation.[2] The theory suggests that microseconds after the Big Bang, the universe expanded at remarkable speed, spreading out from a point into a space some 100 septillion (100,000,000,000,000,000,000,000,000,000) times as large. That supercharged expansion smoothed out any inconsistencies, leaving the nearly-uniform background we see today. According to a recent paper by the California Institute of Technology's Ranga-Ram Chary,[3] "Collision between these regions, if they occur, should leave signatures of anisotropy in the cosmic microwave background." (Anisotropic objects show a different strength when measured from different directions. For example, wood is stronger along the grain than across it.) Some researchers believe they have found just that, providing circumstantial evidence to support the parallel universe or "many-worlds" concept (Fig. 80.1).

Until recently, our tools were not sensitive enough to map the universe's cosmic background radiation in detail. In 2011, researchers using data from NASA's Wilkinson Microwave Anisotropy Probe (WMAP, launched in 2001) looked for places where another universe might have grazed ours, but they found nothing. Then came the European Space Agency's far more sensitive Planck Mission. The Planck spacecraft operated from 2009 to 2013, studying the most ancient light of the universe. And within its vast amounts of data,

[2] Alexei Starobinsky, Alan Guth, and Andrei Linde won the 2014 Kavli Prize for the initial theory.
[3] Spectral Variations of the Sky: Constraints on Alternate Universes; R. Chary; Astrophysical Journal; v.30 Nov 2015; arXiv:1510.00126

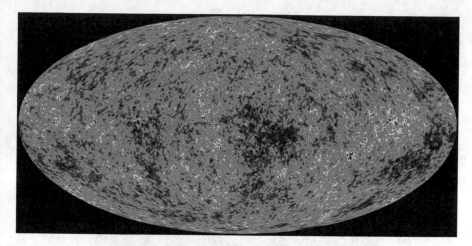

Fig. 80.1 A map of the universe's background radiation charted by NASA's Wilkinson Microwave Anisotropy Probe. Does this (and later, more detailed maps) show sites where our universe has bumped into others? (Courtesy NASA/WMAP Science Team)

astrophysicists have discovered that the background radiation isn't so smooth after all. Ranga-Ram Chary is one of them. Chary sifted through Planck's map of the cosmic background radiation and found three unusual patches of light. In these regions, data suggests that the hydrogen present is up to 4500 times the level expected. It is possible, Chary says, that parallel universes have bumped into ours at those sites, dumping extra material into our own universe.

Physicist Kara Szathmary is also interested in possible sites of "bruising" in the Cosmic Background Radiation, and in particular "the so-called 'Hawking Points'. These points are warmish locations in the very cold area from ancient earlier universes which have had all their mass fall into black holes. Those objects are releasing Hawking Radiation into their former universes." These areas may be remnants of very strange universes, filled with gravitons and photons only. All of their stars have already fallen into black holes, and those black holes are merging with each other, says Szathmary. "Hawking Points are former universes that are in the dying stage." Hawking radiation is emitted as a black hole dissolves over time. These Hawking points are, according to Nobel laureate Roger Penrose and other researchers, remnants of earlier universes. Penrose asserts that the Planck data reveals six separate sites marking massive black holes, ancient universes providing glimpses into other eons. Szathmary adds that, "those universes are ready to go through their next Big Bang once all their black holes evaporate. Talk about a multiverse!" (Fig. 80.2)

Physicists began to seriously suggest the existence of other universes with the rise of inflationary cosmology. While inflation describes how the universe

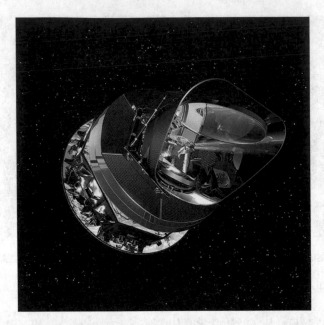

Fig. 80.2 The European Space Agency's Planck observatory has mapped the cosmic background radiation in finer detail than any of its predecessors. (©ESA - AOES Medialab)

behaved just after the Big Bang, the Big Bang itself "smoothed out" any evidence of the moment of its birth. Still, we can peer back in time with our telescopes and other instruments to see conditions in the very young universe. Inflationary cosmology reveals a force that generated the outward rush of the universe. But the power behind this force was so great that many physicists envision many universes emanating from the initial Big Bang. Each is balanced differently, with their own set of physical laws. Ours may be the only one stable enough to enable our kind of life to take hold (Fig. 80.3).

Is it possible that a second Xavier—Xavier Ω in our story—or a second you, exists in such universes? Chary thinks not. In a recent paper, he explains that even if the Many Worlds concept is true, "many other regions beyond our observable Universe would exist with each such region governed by a different set of physical parameters."

The multiverse theory envisions our universe as one of many, like soap bubbles that pop into and out of existence constantly. Some of the bubbles are unstable and short-lived. Others simply collapse back into themselves. Chary says, "Basically, the physical properties in those other universes, if they exist, are so different that the probability that they still exist after the 13.8 billion years of our universe is very remote. In other words, there are many other

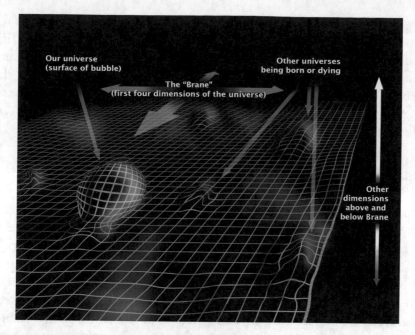

Fig. 80.3 The Universe we experience has four dimensions: width, depth, height, and time. These four dimensions are illustrated here as a two dimensional plane, called the Brane (short for membrane). Perpendicular to our four dimensions are at least six other dimensions (according to String Theory). Our universe (and possibly others) issue from this plane as bubbles. Our own universe is expanding, seen here as the large foreground bubble. We exist only on the surface of the Brane. Outside of that plane are the other dimensions that we cannot—as yet—see. But researchers continue their attempts to confirm other dimensions. (Art by the author and Kara Szathmary)

failed universes to create one like ours. The probability that another one can sustain life or have human life forms is, therefore, infinitesimally small."

Physicist Ethan Siegal agrees. In a 2015 paper,[4] he says, "Even setting aside issues that there may be an infinite number of possible values for fundamental constants, particles and interactions, and even setting aside interpretation issues such as whether the many-worlds-interpretation actually describes our physical reality, the fact of the matter is that …unless inflation has been occurring for a truly infinite amount of time, there are no parallel universes identical to this one." Still, the idea of a Xavier Ω makes a good story, and that is what the novel portion of our book relies upon.

Our bubble universe has a host of variables that, taken as a whole, have led to a stable universe with stars, planets, galaxies, and a perfect blend of dark

---

[4] *Ask Ethan #73: The Multiverse and You*, Medium.com, 2015. https://medium.com/starts-with-a-bang/ask-ethan-73-the-multiverse-and-you-46c9e3c493e2#.xo04du8md

energy (an ever-increasing driving force of the universe's expansion) and dark matter (unseen matter in the universe). As the "friar of ChronoCorp," Todd, asked, "How did our universe end up with such a perfect mix, a finely tuned balance?" Popular physicist Brian Greene puts it this way: "This is a deep question: why are those numbers so finely tuned to allow stars to shine and planets to form when we recognize that if you fiddle with those numbers…any fiddling makes the universe disappear."[5]

Closed Time-Like Curves

Time can be thought of as a straight line lying across the plane of space/time (that imaginary plane that includes our three dimensional universe of height, depth, and width, plus the fourth dimension, time). If a massive object (like a planet or black hole) depresses the fabric of space/time around it, the line will bend as it passes through that "dip." If it bends enough, it becomes a curve, called a closed time-like curve. But to loop back upon itself requires an added feature, a feature that physicists refer to as "frame dragging." Imagine that our massive object, nestled within a dent in the space/time plane, begins to spin. As it does, it twists the plane around itself.

Xavier explains that, "The ring laser swirls empty space like a rotating spoon stirs molasses. It twists the time that we perceive as a linear affair, folding it around into that loop Katya mentioned…This disruption of the space/time continuum is essentially our portal to other locations on the timeline." Is this pure fantasy, or does science allow for it? The mathematician Kurt Gödel, a colleague of Einstein's at Princeton, proved mathematically that travel into the past was at least possible without violating any laws of nature. Indeed, Einstein's Special Theory of Relativity demonstrates backward time travel using faster-than-light travel, and his General Theory of Relativity allows for backward travel using closed time-like curves. Einstein asserted that the speed of light cannot change. This can only be true, he stressed, if time itself changes. Time is affected by speed. The faster you go, the slower time passes. His initial theory of relativity dictated that no matter how fast you go, you cannot go backwards in time. But his later work in special relativity demonstrated that backwards time travel is possible. The reason: time is anchored to space. Gravity is the warping of space, and it is also the warping of time. For example, engineers initially couldn't get car GPS's to sync up with the GPS satellite system, because the GPS receivers in cars ran slower than those on the

[5] YouTube TED Talk presentation "Brian Greene and String Theory". https://www.youtube.com/watch?v=YtdE662eY_M

satellites. Why? Because the cars were under more gravity than the orbiting satellites.

And what about that spoon analogy? General Relativity predicts that a rotating mass drags space/time around it, just as a rotating spoon will twirl molasses in a cup, which is what frame dragging is all about. The phenomenon of frame dragging was recently confirmed in a study[6] of a binary star system with the elegant name of PSR J1141-6545. The system consists of a white dwarf star in orbit around a pulsar. Researchers predicted the times of radio pulses coming from the pulsar, and discovered a "drift" in the orbit of the white dwarf. Their conclusion is that the observations are revealing frame dragging. The rapidly spinning white dwarf is warping and twisting the fabric of space/time around itself.

The Earth itself pulls the fabric of space/time around itself, too, although to a much subtler effect. A study led by Ignazio Ciufolini at the Universit di Lecce[7] used two satellites, LAGEOS and LAGEOS 2, to search for the effect. Using thousands of laser pulses to measure the orbits of the two craft, the researchers were able to monitor how the Earth bent space around itself, affecting the orbits of the satellites.

How does frame dragging relate to closed time-like curves? As our timeline bends into the "depression" around our heavy object, it becomes twisted because of frame dragging. If it twists enough, it could theoretically bend back upon itself. Anyone traveling along that line could follow it back to where it meets itself in the past. In our story, this is accomplished not with a dense black hole (which would also do the trick nicely), but rather with the spiral of lasers. The idea is that instead of using a huge, heavy object, the time-line can be looped by the use of light.

Physicist Ronald Mallet points out[8] that time is moored to space. We know that gravity is essentially the warping of that elastic space/time plane. And gravity also warps time. Models demonstrate that time slows as a traveler approaches the speed of light. Mallett's work shows that if light affects gravity, and gravity affects time, their interrelationship means that light also affects time.

[6] *Lense-Thirring frame dragging induced by a fast-rotating white dwarf in a binary pulsar system* by V. Venkatraman Krishnan, et al., *Science 31*, Jan 2020, Vol 367, Issue 6477

[7] *Satellite Laser-Ranngng as a Probe of Fundamental Physics* by Ignazio Ciufolini, et al., *Scientific Reports 9*, Article number 15881 (2019)

[8] See *Time Traveler: A Scientist's Personal Mission to Make Time Travel a Reality* by Ronald Mallett and Bruce Henderson, Thunder's Mouth Press 2006

Mallett's work has been critiqued in the physics community (that's the way science works), and he has honed his work to answer his colleagues. It is unique work, and has inspired other studies of practical time travel.

Would a closed time-like curve really work? Many physicists point out that Einstein's mathematics allow for it, but physicist Marshall Barnes says,[9] "closed time-like curves are, in reality, nothing more than smoke and mirrors." He goes on to call them "one of the most abused and misunderstood abstractions from Einstein's theory of General Relativity." His paper concludes that, "…in the end, Closed Time-like Curves are more than mathematical errors due to inaccurate interpretations of the nature of time. They're a concept, in and of itself, that not only has very little understanding in the physics community, but isn't even needed in order for time travel to exist."

Barnes suggests that the only solution to the paradoxes of backward time travel is the parallel universe concept. "The moment that any curve attempted to break the barrier between the present and the past, the time traveler would be in the new past copy, *instantly*, discontinuously. There would be no line that just ended up that way. All time travel is *discontinuous*." The question then becomes: can our traveler return to his or her original universe, or do they now permanently inhabit the alternate one which their actions birthed?

Just as observatories like the WMAP charted the background radiation of the Big Bang, Ronald Mallett has a plan for observing frame dragging, and its possible time warping, through a laboratory experiment. Mallett is seeking funding to build a powerful ring laser, a system that shunts a laser into a circulating pattern, in an attempt to twist time into a loop. But it's a big jump from a laboratory laser to a time machine, Professor Michio Kaku told *Closer to Truth* in an interview. "For a time machine, you need the energy of an exploding star." Physicist Kara Szathmary agrees, commenting that, "to achieve time-like motions requires HUGE amounts of energy and accelerated speeds which would be worse than blurring, let alone remaining conscious. So, I cannot imagine humans experiencing that possibility."

Paradoxes

A paradox is a seemingly self-contradictory situation that, when investigated, proves to be true. Time travel is full of them. Xavier outlines a paradox of time travel by mentioning a traveler who goes back to assassinate his own grandfather. Once the deed is done, the time traveler ceases to exist. But since he no

---

[9] *Killing closed time-like curves (for their own good and ours)* by Marshall Barnes

longer exists, his grandfather cannot be assassinated, and so the time traveler is born anyway. This is the famous "Grandfather Paradox," but the paradox can take on more complicated variations. A man and woman meet for dinner. It's a first date, and the punctual man gets there first. He and his girlfriend have a lovely dinner, fall in love, and eventually marry and have children. One of those children becomes the father of a physicist. He tells her stories of what a wonderful man her grandfather was. She hops into a time machine and goes back to that fateful day when the two are supposed to have dinner. She shows up early and tells her grandfather-to-be astonishing things that he will do, and wondrous things about the future. She hides before her grandmother-to-be arrives, but now her grandfather is so flustered that his date finds him distracted and not entertaining at all. They go their separate ways, never to connect and have a family. What happens to the physicist granddaughter?

The grandfather paradox—in all its versions—is classical relativity. But as Marshall Barnes and others suggest, quantum relativity allows this paradox because it posits multiple universes or realities. At the instant a traveler arrives in the past—according to the Many Worlds scenario—the universe splits and the traveler is in a different universe. But the question is this: when they return, do they return to their own, original universe?

Oxford University physicist David Deutsch agrees that if you travel back in time, when you arrive in the past there is a split in the universe, landing you in a parallel one. In that universe, you can do whatever you want without paradoxical consequences. But there is a branch of the universe that you cannot arrive in, and that branch is the reality in which you existed in the first place. Says physicist Ron Mallett, "For me, I tend to like that particular interpretation because it resolves paradoxes. Which of these things really plays out? It turns out that reality may just be more plastic than we think. We might be able to go back and alter our universe. And even though it leads to a paradox, the universe may, in fact, contain paradoxes. We don't know what the universe can actually do. A lot of physicists seem to like the parallel universe version because it easily resolves these paradoxes."

Another solution to the apparent paradoxes of travel into the past, mentioned by our character Xavier, is that the timeline self-corrects, with the outcome remaining the same no matter what the traveler does. This scenario seems to be confirmed by recent work, including a paper[10] from the University of Queensland. Work by Germain Tobar and Fabio Costa indicates that the traveler would still have free will within their immediate "surroundings" of

[10] *Reversible dynamics with closed time-like curves and freedom of choice*, by Germain Tobar and Fabio Costa (IOP Publishing Ltd., September 21, 2020)

time and space, but the timeline would adjust so that whatever actions they took would eventually be wiped out. In an interview with Queensland University's news service, Germain Tobar used the example of an attempt to prevent the coronavirus by protecting the person at its source. In stopping patient zero from becoming infected, he said, "in doing so, you could catch the virus and become patient zero, or someone else would…No matter what you did, the salient events would just recalibrate around you."

Time Travel in Our Culture

Ronald Mallett's interest in time travel was piqued by the classic H.G. Wells novel *The Time Machine*. Wells' book was one of many. Even before Wells, in 1884, Edwin A. Abbott penned *Flatland: A Romance of Many Dimensions*. In it, he describes a two dimensional world inhabited by circles, triangles and other geometric figures. He also includes other universes like Pointland and Lineland. DC and Marvel Comics are rife with sagas taking place in parallel universes. Steven King's *Dark Tower* series, C.S. Lewis' *Chronicles of Narnia*, and Michael Crichton's *Timeline* all include multiverse or parallel universe concepts.

Parallel or multiple universes have been explored in other media as well. TV and movies have taken up the gauntlet, with a long parade of titles. The Time Lords of *Dr. Who*, the personal interactions in *Continuum*, and Rod Serling's *Twilight Zone* touch upon multiple dimensions or universes. One of the earliest television series to explore the concept of parallel universes was the original *Star Trek*. In the episode *The Alternative Factor*, an extra-dimensional traveler named Lazarus (played by Robert Brown) battles to prevent his identical twin from obliterating both realities. In another episode entitled *Mirror, Mirror*, members of the Enterprise crew accidentally beam into a parallel universe where their counterparts are militaristic. And speaking of counterparts, the excellent Starz network series *Counterpart* starring J. K. Simmons explores the ramifications of a "tear" in the universe that appears in Berlin, linking two different but parallel universes. D.C. Comic's *Legends of Tomorrow* (with Kevin Mock) rounds out our short list of TV series here, but there are many others.

More generally, movies about time travel abound. A few highlights include *Twelve Monkeys* (Bruce Willis, Brad Pitt, Madeleine Stowe), *Time Lapse* (Danielle Panabaker, Matt O'Leary, George Finn), *Edge of Tomorrow* (Tom Cruise, Emily Blunt), *Source Code* (Jake Gyllenhaal, Michelle Monaghan, Jeffrey Wright), *Predestination* (Ethan Hawke, Sarah Snook), *Primer* (Shane

Carruth, David Sullivan), *Donnie Darko* (Jake Gyllenhaal), *Interstellar* (Matthew McConaughey, Jessica Chastain, Anne Hathaway), *About Time* (Domhnall Gleeson, Rachel McAdams), *Planet of the Apes* (Charleton Heston), *Back to the Future* (Michael J. Fox, Christopher Lloyd), and the *Terminator* series (Arnold Schwarzenegger, Linda Hamilton). The long parade of literature, movies and other media demonstrate just how deeply the concept of time travel and, more recently, the multiverse has infused modern culture.

Dinosaurs and Paleontology

The Real Benjamin Waterhouse Hawkins (Fig. 80.4)

There exists a long and rich history of the melding of art and science. Thomas Moran and Albert Bierstadt accompanied explorers of the western US (where Native Americans had been living for millennia). Their paintings helped to convince the US Congress to establish its first three national parks, Yellowstone, Sequoia, and Yosemite. Nearly a century later, Chesley Bonestell depicted human exploration of space, decades before history caught up. Like the works of Moran and Bierstadt before him, Bonestell's paintings inspired a generation of new scientists and explorers. Similarly, the masterpieces of Benjamin Waterhouse Hawkins—paintings and sculptures—inspired a generation of researchers in the field of paleontology, and inspired others to enjoy the natural sciences.

Hawkins' pioneering work seems outdated to the modern eye, but his depictions mark one of the first serious attempts to render accurate dinosaurs. His reconstructions set a pattern that would last a century. In fact, a cursory glance at any book from the 1950s reveals similar versions of dinosaurs, lumbering, lizard-giants wandering across Mesozoic landscapes. Many drag massive tails, and wear scaly coverings over obese bodies. Some of these portrayals of slow-moving, cold-blooded behemoths—many of them crafted by the famous wildlife master-painter Charles R. Knight—changed little from those of Benjamin Waterhouse Hawkins' masterpieces.[11]

Born in Bloomsbury, London in 1807, Hawkins became a wildlife artist and sculptor. He contributed paintings and drawings to the publication of

---

[11] Although Knight's 1897 painting of two battling *Dryptosaurus* (known at the time as *Laelaps*) anticipated the modern view of active dinosaurs by nearly a century.

Fig. 80.4 Waterhouse Hawkins reconstructed the *Hylaeosaurus* as a heavy lizard-like creature (top). This sketch was done in preparation for his giant sculptures at the Crystal Palace in London. Recent reconstructions are more streamlined, with the creature holding its tail erect. *Hylaeosaurus* was the first ankylosaur (armored dinosaur) discovered. Some spines may have protruded from the back, but the majority likely extended from the sides

(above: public domain, Wikipedia Commons: https://upload.wikimedia.org/wikipedia/commons/7/78/Hylaeosaurus.jpg

below: art by the author)

The Zoology of the Voyage of the HMS Beagle, and went on to create sculptures for the Royal Academy.

Hawkins was engaged as assistant superintendent of the 1851 "Great Exhibition of the Works of Industry of All Nations," an international fair held in London's Hyde Park. Queen Victoria and Prince Albert sponsored the event. The centerpiece of the exhibition was the Crystal Palace, a spectacular glass building draped upon a skeleton of iron. Three times the size of Saint Paul's Cathedral, the soaring building's central rotunda rose some 130 feet (40 meters) high. Its glistening skin encompassed 900,000 square feet (84,000 sq. m) of glass. The following year, the Palace was moved, piecemeal, to Sydenham. The statues of Waterhouse Hawkins, the first life-sized dinosaur sculptures in the world, were destined to populate the grounds surrounding the beautiful edifice at its new site.

Hawkins crafted 33 statues for the event, and they were large. Hawkins and Richard Owen, the paleontologist who advised Hawkins, held a dinner party on New Year's Eve *inside* the mold for the *Iguanodon* statue. The sculpture mold was large enough to accommodate a table and chairs, with Owen at the head of the table, inside the Iguanodon's head.[12]

Often Hawkins had little fossil evidence to work with, but he had a formidable ally in the British paleontologist Richard Owen.[13] Some prehistoric fossils offered Owen and Hawkins only fragments of bone and claw. One such creature, the flagship statue of the 33 figures done for the project, was the duck-billed hadrosaur *Iguanodon*. Hawkins had only several teeth and a "horn" of an Iguanodon available for his reconstructions. First described by Gideon and Mary Ann Mantell,[14] the teeth resembled those of an iguana, but at twenty times the size. Later work revealed the associated horn to be a thumb spike instead of a rhino-like nasal projection. Paleontologists came to realize that the *Iguanodons*, like all hadrosaurs, often held their tails erect, and could occasionally stand on all fours or bipedally. This has been confirmed through studies of skeletal structure, and also through trackways, which often lack tail impressions completely. However, many tracksites preserve evidence of hadrosaurs walking on all fours, walking on their hind feet, or turning. As the

---

[12] Many of these statues still stand today near the site of the original Crystal Palace. Today they can be found in the London suburb of Bromley at the Crystal Palace Park.

[13] Owen was the paleontologist who first coined the term *Dinosaur*.

[14] There is some confusion as to which of the Mantells first discovered the teeth. Although the popular accounts say that Mary Ann discovered the first fossils while wandering a roadway outside of a patient of Gideon's, but there seems to be no evidence that she actually accompanied her husband during his doctor visits. In 1851, Gideon Mantell said that he found them, although he had asserted in 1827 that she had. Mary Ann was the one to christen the creature Iguanodon.

creature turns, there are occasionally impressions where the tail just skimmed the ground (Fig. 80.5).

After going on a lecture circuit, Hawkins made his first trip to America in 1868. There, he assembled the first fully articulated dinosaur skeleton of a hadrosaur that had been discovered in a New Jersey clay pit. Hawkins went on to give a series of popular lectures on natural history and began work on the Paleozoic Museum which was planned for New York's Central Park.

Fig. 80.5 The studio of Waterhouse Hawkins in Sydenham, outside of London. (Public domain: https://commons.wikimedia.org/wiki/File:Sydenham_studio.jpg)

Four decades before, he had married his first wife, Mary, and had several children before bigamously marrying his second wife, Louisa, ten years later. After the disastrous destruction of his work in New York, he returned to England in 1874, but by that time Louisa had learned that their marriage was not legal. Hawkins returned to America within the year, possibly due to what he termed as a "climax of domestic troubles." He continued his work at Princeton University, painting ancient life reconstructions until his final return to England 1878, where he cared for his ailing former wife Mary (by then, he had divorced Mary and married Louisa, but the marriage was likely only a legal one to secure the reputation of his children). He died of a stroke in 1894, leaving behind a legacy of historic paintings and sculptures, inspiring people of all ages on both sides of the Atlantic and beyond.

The story of Boss Tweed and the destruction of Hawkins' work in Central Park is true, sadly. Many searches have been undertaken to find remnants of his sculptures beneath the lawns and fields of the Park's southern areas, to no avail. Several major construction projects, including excavations for subway lines, have yielded no artifacts. Some sources assert that shortly after the event, the statue fragments were dug up and found to be of no value, so they were discarded. Others say that Hawkins himself may have recovered at least a few pieces, with their ultimate fate unknown. Still others say that they were crushed up and used for paved paths in the Park, where even today, people walk across the masterpieces of Benjamin Waterhouse Hawkins.

Dinosaur Scales, Feathers, and Posture

In the early days of paleontology, many of the dinosaur skeletons appeared to be lizard-like in form, and restorations depicted them in a tripod pose, standing on back legs and using the tail for stability against the ground. But as more dinosaurs were uncovered and techniques like collection, preparation and analysis improved, scientists realized that many of the prehistoric creatures were built like today's birds. Far from lumbering behemoths, dinosaurs were agile and quick.

But what of their covering? Did the long-necked apatosaurs wear the wrinkled skin of the rhino or elephant? Did *Tyrannosaurus rex* sport a scaly lizard's exterior? Were the horned ceratopsians blanketed by the armor of the crocodile? Skin impressions of prehistoric creatures are hard to come by, as skin is rarely preserved in the fossilization process. But its imprint is sometimes left behind, and we are beginning to see patterns in the various types of dinosaurs. Many tyrannosaurids like *Albertosaurus, Daspletosaurus* and *Tyrannosaurus rex*, had scaly, reptilian skin. Samples have been found from tails, flanks, the neck and pelvis of various tyrannosaurids, all with scale patterns. But some Late Cretaceous tyrannosaurids found in China possess feathered plumage.[15] One of them, an ancestor of *Tyrannosaurus rex* called *Yutyrannus huali*, had what appears to be a blanket of fluffy down along its entire 30-foot (10-m) length.

Of the reptilian skin samples, scales come in different sizes and shapes, varying from round to hexagonal, and from smooth to raised. Modern reptiles show such variations, and those variations often relate to color changes. Some dinosaurs (most notably the duck-billed hadrosaurs) were "mummified" before the process of fossilization. Their dried skin left impressions in the surrounding rock, formed in the burial sediments of sand prior to the skin decaying and the sediment lithifying (turning to stone). Some scales have circular arrangements, while others vary in stripes. Different-sized scale patterns found on a mummified *Edmontosaurus* nicknamed Dakota show stripes across the arms, for example. A study[16] by Phil R. Bell of the University of Alberta compared skin impressions of several *Saurolophus* specimens. His work indicates likely mottling on the tail of one, while another has vertical alternating patterns of large and fine scales, suggesting zebra-like stripes (Fig. 80.6).

Fig. 80.6 The scales of this Triceratops fossil skin sample vary in size and shape, perhaps indicating differences in coloration. (Courtesy Kent Bush, the *Rapid City Journal*)

[15] Xu X, Norell MA, Kuang X, Wang X, Zhao Q, Jia C. 2004Basal tyrannosauroids from China and evidence for protofeathers in tyrannosauroids. *Nature* **431**, 680–684.

[16] Standardized Terminology and Potential Taxonomic Utility for Hadrosaurid Skin Impressions: A Case Study for *Saurolophus* from Canada and Mongolia by Phil R. Bell, February 3, 2012, PLOS ONE; https://doi.org/10.1371/journal.pone.0031295

The concept of feathered dinosaurs has been around for a long time. In 1868, Thomas Henry Huxley proposed a link between dinosaurs and modern birds. Huxley rendered a feathered Compsognathus[17] for his lecture on bird evolution in 1876. Huxley's work was inspired by an 1861 discovery in a Bavarian limestone quarry. There, workers split open a slab of limestone to find a beautifully preserved creature, a cross between a lizard and a bird. Other similar specimens were found, and many shared one thing in common: preserved impressions of feathers. Their skeletons also had bird-like wishbones, but their tails seemed to be more reptilian, as did their toothed mouths. These were christened *Archaeopteryx*, the first true birds ever found. They came from the time of the dinosaurs, and caused Huxley to wonder if true dinosaurs were similarly adorned.

But with no fossil evidence of feathered dinosaurs, the idea faded from popularity. Fast forward to 1969, when paleontologist John Ostrom described the theropod Deinonychus antirrhopus, pointing out its skeletal similarities to birds. Ostrom and his spunky student Robert Bakker became great advocates of a dinosaur/bird connection, with all its ramifications. Ostrom and Bakker argued that dinosaurs were energetic and agile, possibly warm-blooded, traveling in herds, and caring for their young.[18] By the 1970s, many researchers had come to the same conclusion, and paleontological reconstructions often showed dinosaurs with feathers. In 1996, Canadian paleontologist Philip Currie showed a snapshot of what was probably the first feathered dinosaur to be discovered, this one in China. Scientists discovered increasing numbers of well-preserved dinosaur fossils with associated soft tissue like organs, skin, and feathers. Many come from rich fossil beds in Liaoning, China. Chinese discoveries include such feathered dinosaurs as *Sinosauropteryx*, *Velociraptor*, and *Microraptor*.

Dinosaur skin and feather impressions had not been identified at the time of Waterhouse Hawkins' work. He and Richard Owen had far less information to work with than today's paleoartists, who have the advantage of the many spectacular discoveries over time (Fig. 80.7).

---

[17] No feathers have been found on *Compsognathus* fossils, but dinosaurs similar in structure do have preserved feathers.

[18] For more on this, see the groundbreaking book *The Dinosaur Heresies* by Robert Bakker (Citadel, 2001)

Fig. 80.7 Cast of the feathered dinosaur Caudipteryx zoui, found in Liaoning, China. The elegant therapod was the size of a peacock. The cast is on display in the Houston Museum of Natural Science (Daderot, CC0, via Wikimedia Commons: https://en.wikipedia.org/wiki/Caudipteryx#/media/File:Caudipteryx_zoui,_feathered_dinosaur_plate,_Early_Cretaceous,_Yixian_Formation,_Liaoning,_China_-_Houston_Museum_of_Natural_Science_-_DSC01866.JPG)

Ancient Thera and the Minoans

Thera and Atlantis

Science often interacts—and sometimes conflicts—with the "softer" sciences of history and cultural mythology. A prime example of the healthy confluence of anthropology, archeology, and mythology lies in the legends of Atlantis, first put forth in the writings of Plato.[19] All three disciplines lend insights into the others.

---

[19] For more, see *Santorini: a guide to the island and its archaeological treasures* by Christos Doumas, pp. 62–3 (Ekdotike Athenon, originally published 1980)

Plato's accounts appear in the dialogs of Timaeus and Kritias, both written about 360 BCE. Plato traced his narrative back to Kritias and his great grandfather. His great grandfather in turn heard it from his father, Dropidas. Solon, an ancient legislator, passed it to Dropidas' father from the Athenian Sage and the priests of Sais in 590 B.C., when Solon traveled to Egypt. Plato's account from the ancient Egyptian priests asserted that the kings of Atlantis had dominion over a vast area. Atlantis itself was said to be a continent the size of North Africa and Asia Minor combined, and Atlantean influence spread to parts of Libya and Egypt, and to regions in Europe as far as Northern Italy.

Plato describes Atlantis as having buildings made of red and black stone, paralleling the color of volcanic pumice on the modern island of Santorini, site of the ancient capital of the Minoans, Thera. His description of concentric harbors has been interpreted by some as an echo of a collapsed volcanic caldera (central crater) which, indeed, Thera was. His portrayal of heated spring-fed fountains is a tempting counterpart to the kind of hydrothermal springs seen on Thera/Santorini today. And Plato's assertion of the Atlantean influence over a wide region, and its end by sinking into the sea, provide clear parallels to the true account of Thera and Crete during the Minoan heyday.

Although Crete served as the central hub of the Minoans, the Egyptian priests may have confused the two locations, assuming that the destruction and "sinking" of Thera represented the same location and substance as Crete. Plato's text relates that Atlantis sunk into the sea 9000 years before Solon. But as with many ancient accounts, numbers are often exaggerated. Scholars have corrected the timeline of the accounts, suggesting that if the original number was 900 years before Solon (off by a factor of ten), the date fits quite well with the estimated time of the Theran eruption. Plato's ancient texts may actually describe not a fanciful advanced civilization, but rather the actual Minoan empire and its fall. Plato's account serves as a cautionary tale: Atlantis was a utopian state that prospered and flourished for as long as its people honored and enforced the law and worshipped their gods. But when its citizens became arrogant and turned their backs on law and religion, the gods obliterated the entire continent.

Ancient Utopia?

It is important to remember that Plato's dialog was not a scientific treatise, but rather a moral tale. And although Thera and Crete may have served as the kernel for the Atlantis narrative, they were not a continent. Still, some elements of the Minoan society and culture seem to at least parallel Plato's narrative. The neopalatial period in which our time travelers explore occurred throughout a roughly 260-year span between 1750 and 1490 B.C. As Mila van Dijk pointed out, it was a time of prosperity and expansion for the Minoan empire, an empire ruled not by military force but by trade and economy. At the heart of the Minoan empire was the island of Crete with its expansive cities and palaces. Crete's ancient wonders include the Palace at Knossos, a sprawling multi-storied set of structures. The legend of the labyrinth, with the Minotaur at its center, may have come from the labyrinthine causeways leading to Knossos' central courtyards. The bull was an important religious symbol in the Minoan pantheon, perhaps tying in with later myths of the Minotaur, a creature half human and half bull. Some accounts assert that bulls were kept in the central courtyard of the king's palace. Plato also described the importance of the bull in the ancient Aegean, and detailed the bull-jumping gymnasts of the Minoans.

Knossos represents the zenith of preserved Minoan architecture. Interior rooms were illuminated from above with the use of light wells. Many buildings were multi-storied, and roofs were utilized as yet another level for practical use. Columns line the long causeways, many tapered so that their capitals are larger than their bases. Columns and other architectural elements were often painted brick-red, while blue, white, or red striping ran along the rooflines. Stylized bull horns became a key element of edifices—repeated frequently—especially in sacred places (Fig. 80.8).

Thera serves as a target for the time travelers in our novel. The maritime location was an important seaport for the Minoan lands and surrounding islands of the Cyclades and the northern Aegean. Beyond, evidence reveals contact with the Egyptians and civilizations in Asia Minor. Indications of sea trade dating back to 3000 B.C. (a period known as Early Minoan I) comes from Minoan artifacts found at distant locales, along with the remains of several ships discovered at the ancient Minoan harbor adjacent to the great palace at Knossos, at a site called Poros. Some cargo attests to the fact that the Minoans were familiar with metallurgy. In addition to technology like bellows and crucibles, many copper ingots and clay molds set sail in the bellies of Minoan ships. Bronze and silver daggers have been unearthed in tombs as well. Added to the list of trade goods were obsidian (black volcanic glass), grapes, grains, and saffron.

Today, the island of Thera is known as Santorini. Santorini is the remnant of an immense cone-shaped stratovolcano. All that is left of that prehistoric mountain is a crescent-shaped main island to the east (Santorini itself), with two smaller islands called Therasia and Aspronisi completing the ancient caldera ring. Two even smaller islands in the center, Nea Kameni and Palea

Fig. 80.8 Entrance to the Grand Staircase within the great palace complex of Knossos. Note the tapered columns, a feature common in Minoan architecture. (Photo by the author)

Kameni form the tip of the still-active submarine volcano—all that is left of the great prehistoric volcanic cone rising out of the sea.

Much of what we know of ancient Thera comes from the ruins of a city buried beneath Thera's ash. In a strange twist of history, the ruins were discovered because of a construction project 1800 miles (2900 km) away. In 1859, construction began on the famous Suez Canal, a waterway designed to connect the Mediterranean and Red Seas through Egypt's Isthmus of Suez. Because of its volcanic history, Santorini became a primary source of pozzolan, volcanic ash used in water-resistant cement. But while excavating the volcanic ash, workers uncovered the remains of the 3000-year-old seaport of Akrotiri, beautifully preserved beneath the ash. Akrotiri has been called the Pompeii of the Aegean. Its ruins contain storage vessels, ceramic pots, and sculptures, gold figurines, multi-storied apartments, and exquisite murals.

One difference between Italy's Pompeii and Santorini's Akrotiri is bodies. None have been found in Akrotiri, leading experts to suggest that the inhabitants had plenty of time and warning in which to escape the eruption in their ocean-going vessels (Fig. 80.9).

Fig. 80.9 A row of pithoi vats preserved in Akrotiri, probably against the wall of a storehouse or market. Note the beautiful designs painted on the exteriors. Residue from original contents, like grain or wine, can still be found inside the vats, encased within Thera's ash. (Photo by the author)

Minoan Plumbing

The concept of a steam-driven secret room is firmly rooted in the realm of fiction. We have absolutely no evidence of steampunk Minoans. But the ancient Minoans did have an interest in steam and piping. In fact, Crete and Thera enjoyed the luxury of indoor heating. The Minoan engineers accomplished hypocaust (central) heating by the use of terra cotta pipes carrying hot air or water beneath the floors. Architecture on Santorini, Crete and adjacent islands shows that apartments had access to spring-fed heated water. One such example is the "Queen's bathroom" at Knossos, where sub-floor pipes carry heat to a painted terra cotta bathtub.

A dual set of pipes in many structures suggests the use of both hot and cold water. Homes, buildings, and public baths used terra cotta pipes beneath the floors. In the hot summers, the flow could be shunted to shut the heat off.

Ancient city planners also used a system of pipes for a sophisticated sewer system. Rainwater draining from roofs and gutters lining town squares, along with the overflow from various cisterns, flowed down through drains into buried clay pipes. These pipes had quite modern joints, with the narrow end of one pipe fitted into a wider end on the next. Similar drainage systems have been found in the Minoan settlements of Gournia, Vathypetro, Tilissos, Malia and others.

The Minoans enjoyed the technology of indoor flush toilets some 3200 years before the famous European inventions of Thomas Crapper. The earliest flush toilet yet discovered resides in the palace of Knossos on Crete. The toilet was screened off for privacy. An indentation in the wall led to the seat, and could be flushed by water from cisterns above. Another indoor toilet is found in the remains of Akrotiri. Like other Akrotiri treasures, the commode was encased in volcanic ash, preserving it and its surroundings beautifully (Fig. 80.10).

Also preserved in Akrotiri's ash—as well as in the ruins of other Minoan sites—were fragments of many murals. These ancient masterpieces depict everything from Minoan gymnasts jumping from bulls to fleets of ships. Naturalistic paintings render idyllic landscapes and seascapes inhabited by monkeys, birds, dolphins, and antelope.

Some of the murals are stylized, while others show carefully rendered portraits. One such image has been called Le Parisienne, or the Minoan Lady. It was probably painted on the wall of the Sanctuary Hall (or on a vase left near it) in the front atrium of the palace of Knossos, although its original location is uncertain. It portrays a young woman wearing a sacral knot at the nape of her neck, indicating that she is an influential woman, perhaps a priestess or goddess. Archeologist Edmond Pottier (1855–1934) nicknamed her "Le

Fig. 80.10 Ancient WC: a sculpted chair set against the wall served as a toilet. Water was piped down the indented wall behind and flowed through the slot in the center of the seat into pipes leading to the city sewers. (Photo by the author)

Parisienne" because she reminded him of contemporary women in Paris. Like other women depicted in the Minoan masterpieces, she wears black eyeliner to emphasize her eyes and rich red lipstick on her mouth. Her white skin provides contrast with black curly hair, something common in Minoan depictions of women. No wonder she reminded our character Todd of the lovely Minoan priestess he met in his time travels (Fig. 80.11).

Another significant mural series is the Frieze of the Fleet in the West House, also known as the Admiral's apartment. This spectacular panorama displays a departing fleet of large ships carrying warriors or dignitaries. Festive banners and garlands imply the latter, as the entire mural seems to be a celebration. People bid goodbye to the sailors from their rooftops and windows. The

Fig. 80.11 The three-millennia-old painting of the woman known as Le Parisienne. Note the makeup on eyes and mouth, as well as the sacral knot at the base of the neck. In Minoan murals, women are usually depicted as white-skinned, while men have rich brown skin. (https://commons.wikimedia.org/wiki/File:The_Parisian,_fresco,_Knossos,_Greece.jpg)

painting depicts Minoan houses, multi-storied buildings. A similar scene shows the fleet's return. The mural has provided a rich source of insight into Minoan culture, clothing, architecture, and shipbuilding. The largest of ships had rows of oars in addition to a large square sail. A 16-m(~50-foot)-tall oak mast slid into a reinforced square for easy removal and installation. Sturdy hemp ropes held it in place, but the single square sail could move freely around the mast to catch the wind. The ships had an open-air cabin with a roof, perhaps made of cloth, to shelter the ship's captain or dignitaries. Some canopies appear to have run nearly the length of the entire vessel.

The Antikythera Mechanism

Katya and Todd discover a complex bronze gearbox serving as an analog computer. While such a device—known as the Antikythera Mechanism—*has* been found, nothing like it seems to have existed as early as our action takes place. Research into the Antikythera Mechanism points to an origin in the second or late first century B.C. Our action takes place over a millennium earlier, so there is no direct link between ancient Thera and the Antikythera Mechanism (like our steam-powered room in Kitane's temple, that's what science fiction is for). Still, the small clockwork device is important in ancient Aegean studies, as it is the first known analog computer. Its thirty interlocking gears forecast the positions of the Sun and Moon, lunar phases, eclipses, and planetary movements.

The device was discovered within a Roman Greek shipwreck off the coast of the island of Antikythera. In 1900, sponge divers spotted a bronze hand protruding from the seafloor, and the next winter Greek authorities sifted through the surroundings. Marine archeologists have uncovered over 200 amphorae (vases with pointed bases), sculptures, and more ordinary objects like oil lamps and nails. Among the items salvaged was the unique bronze mechanism, about the size of a shoebox, broken into three main pieces.[20]

Many of the amphorae are in a style originating in Rhodes. The doomed ship may have been on its way to Greece or Rome from Rhodes. In ancient times, trade routes brought vessels from Italy to Asia Minor. Shipping lanes passed the rugged coastlines of Crete, Santorini and smaller islands, one of which was Antikythera. The Roman shipwreck has been dated to about 70 B.C.

CT scans of the device show engraved labels written in Koine Greek. The main text is a 3500-word instruction booklet. Researchers, including

---

[20] The main portion of the device actually fractured into three pieces shortly after it was removed from seawater. It was submerged in water again to preserve it. Over 80 fragments have been discovered to date.

University College London professor Tony Freeth, compared numbers in the text to the number of teeth in the gears and have been able to reconstruct working models of the mechanism. On its face is a large, clock-like dial showing the 365 days of the year in the structure used on the Egyptian solar calendar. An inner circle displays 12 signs of the zodiac. By rotating a crank on the side of the device's wooden frame, two arms indicated the exact positions of the Sun and Moon. One of the arms was equipped with a spinning two-toned sphere that showed the phases of the Moon. On the opposite side of the box, two dials reflected the repeating lunar and solar cycles in the course of 19 years (the metonic cycle), and the movements of some planets. Inscriptions on the dials reveal that several arms were topped with colored balls, including red for Mars and gold for the Sun. The second dial was arranged in a spiral and charted lunar and solar eclipses. All three dials interlocked, so that specific eclipses predicted on the reverse side would be reflected by a specific date on the front. The device was also keyed to mark the sites of upcoming Olympic games.[21]

Several ancient writers, including Cicero, refer to similar devices. According to Cicero, the philosopher Posidonius crafted a model of the cosmos in the first century B.C. Archimedes was rumored to have built a bronze apparatus in the third century B.C., and the Rhodes mathematician/astronomer Hipparchus (~140 B.C.) fused Babylonian and Greek number theory, resulting in the kind of calculations needed for the Antikythera mechanism. Rhodes is a good candidate city for another reason: the mechanism was able to indicate the dates of a fairly minor athletic competition held annually in Rhodes. In 1970, a team led by Jacques Cousteau discovered coins at the shipwreck which date between 67 and 76 B.C. This discovery led to a second theory, which proposed that the device had its origins in the Greek city of Pergamum. Pergamum hosted a huge library, second only in scope to the great Library at Alexandria.

In 212 B.C. the Romans conquered Syracuse. Accounts report that the general in charge, General Marcus Claudius Marcellus, returned to Rome with a mechanism designed by Archimedes (~287–212 B.C.) that demonstrated the movements of the Sun, Moon and the five planets known to the ancients.

A 2017 study posits that an earlier prototype was built in Rhodes, but that the later version was updated and modified for a wealthy patron in Epirus (northern Greece). The commissioned masterpiece never made it to its owner.

---

[21] Freeth, Tony; Jones, Alexander; Steele, John M; Bitsakis, Yanis, *Calendars with Olympiad display and eclipse prediction on the Antikythera Mechanism* (Nature 454, July 31, 2008)

Nothing would match its complex workings until the earliest clockmakers of Europe, who began fashioning their timepieces in the fourteenth century A.D. The artifact is a tribute to the creativity and sophistication of ancient civilizations across the Mediterranean.

The Volcano Before

Several theories have been put forth concerning how Thera/Santorini may have appeared in the time leading up to the cataclysmic explosion near the end of the Minoan reign. The cone shape of the volcano collapsed and eroded away in a series of prehistoric geologic events, leaving an outer ring long before the Minoan Civilization arose over 5000 years ago. Today, Santorini's crescent is all that's left of the volcano that wreaked havoc throughout the ancient Aegean. But a millennia and a half before the common era, when that havoc occurred, the island's actual layout is a mystery.

What did that island look like? In Fig. 80.12, the top view shows Santorini today, exaggerated in the vertical dimension. North is to the upper left. The center and bottom views portray two possible configurations for the volcano in about 1600 B.C.

The Theran cataclysm may have rivaled the eruption of Tambora, Indonesia in 1815. That eruption was the largest in recorded history. It is clear that flows from Thera's explosion, or the floods that rushed into its empty caldera afterward, triggered apocalyptic tidal waves throughout the Mediterranean and beyond. Excavations on Crete, Israel, and Egypt confirm that meters-high tsunamis caused damage across the entire region.

New work[22] by a team of Greek geologists shows that the caldera may have been landlocked within the island at the time of its collapse. The tidal waves, they conclude, came earlier, thrust out by avalanches of super-heated ash (pyroclastic flow), rock, and molten lava blasting into the ocean. Their work indicates that sea water flooded the caldera after its tsunami-instigating eruption. An undersea canyon links the caldera with the sea to the north. This submarine channel was initially blocked by an ash/lava dam emplaced by debris from the eruption. The dam eventually gave way, releasing ocean water into the dry bed of the caldera. But by then, they assert, the tsunamis had come and gone.

For some time, the eruption was dated at 1450 B.C., coinciding with the collapse of the Minoan culture. But other evidence has come to light.

---

[22] *Post-eruptive flooding of Santorini caldera and implications for tsunami generation*, by P. Nomikou, et al. (Nature Communications, 08 November 2016)

Fig. 80.12 Top: A modern view of Santorini, vertically enhanced. The major city of Fira with its airport is at center, and the ancient ruins of the Minoan village Akrotiri are seen at lower right. The town of Oia is toward the upper left. The middle image displays one possible pre-eruption Thera. In this scenario, the entire northern arc of the

(continued)

Comparative data from other Aegean and Mediterranean sites and from Egypt at the time show that a more likely date is 1530 B.C. A team of Greek researchers estimate an even earlier date of 1610 B.C., thanks to the discovery of buried olive branches. The ash-encased wood enabled the group to date the ashfall—and the trees—at the earlier date using C-14 testing. The 1610 B.C. date also bolsters dendrological studies of buried trees, along with a third line of inquiry, ice core analysis. An "acid signal" in arctic cores, an indicator of volcanic fallout on the ice, dates to the same period.

Ancient Outside Sources

Plato's writings are not the only accounts that may reflect the events surrounding the eruption of Thera. According to a paper by J. Alexander MacGillivray,[23] Egyptian inscriptions may clue us in on the time of year and the specific year of the eruption. After her succession to the throne, Pharaoh Hatshepsut engraved a dedication to her accomplishments in Middle Egypt. The narrative is emblazoned on the rock-cut walls of the temple to the goddess Pakhet. The text refers to Hatshepsut sending braziers to her subjects, who had been driven into temples by total darkness and furious storms. Archeologist Hans Goedicke interprets the description as one resulting from the Minoan eruption. Tephra (compacted volcanic ash) has been found at several sites in the Nile delta, along with a string of locations on the Levantine coast. Goedicke[24] also cites another inscription from the reign of Thutmose III, Hatshepsut's successor, which records the event as follows: "there was no exit from the palace by the space of nine days. Now these days were in violence and tempest: none, whether god or man, could see the face of his fellow." Other inscriptions tell of Hatshepsut giving aid to refugees who inundated Middle Egypt from the

Fig. 80.12 (Continued) caldera rim was intact just before the eruption. The volcano is a steep-sided cone surrounded by a lagoon. Only one harbor would have existed, giving access to the sea toward the southwest between Aspronisi and the Akrotiri peninsula. Akrotiri is a bustling village at lower right (based on model by Friedrich et al. 1988). The bottom view shows another possible layout, with breaks in the outer ring at both the southwest and north. Some researchers assert that a cone-shaped volcano in the center of the island would not be large enough to explain the massive explosion that ensued. Instead they suggest that a super-volcano at center may have completely bridged the rim of the caldera, creating two separate bays. (Based on models by Druitt & Francaviglia, 1992; art by the author)

[23] *Thera, Hatshepsut, and the Keftiu* by J. Alexander MacGillivray, from *Time's Up! Dating the Minoan eruption of Santorini, Acts of the Minoan Eruption Chronology Workshop*, edited by David A. Warburton, 2007.
[24] *The Speos Artemidos Inscription of Hatshepsut and Related Discussions* by Hans Goedicke. Halgo, 2004

region of the Nile delta, perhaps as a result of tidal waves. At the same time, she praises her gods for sparing southern Egypt from the dark cloud, and the diverting the floods into the Red Sea.

The Timeline

With analysis of tree rings, carbon dating, geology, and written accounts, some archeologists propose an updated and detailed timeline of the Theran eruption.

1. The action began with a major earthquake. This tremor was strong enough to cause some structural damage, but most buildings in Akrotiri likely remained standing. Evidence shows that inhabitants moved furniture into the streets and began to repair damage before the next event occurred. Damage to the palace of Knossos on nearby Crete was more severe, enough so that the palace needed to be rebuilt.
2. Shortly after the earthquake, a plume of tephra rose from the islands or land bridge at the center of Thera. The cloud of ash drifted southeast, blanketing Akrotiri and surroundings with a fine coating of ash. This may have been the event that triggered the mass exodus from Thera, as inhabitants would have had a short period to collect valuables and escape to Crete and other islands.
3. Thera announced its primary eruption with the explosion off several cubic kilometers of pumice and ash in a near-vertical column. The column of gasified stone rose to roughly 38 km, and circled the entire northern hemisphere of the planet. The raining rock and ash buried Akrotiri with seven meters of searing detritus, preserving the town for modern archeologists to find over three thousand years later. This major eruption may have lasted up to eight hours.
4. The volcano's central vent expanded, and seawater breached fractures in its wall. The salt water mixed with the magma, sparking explosive ejections of boulders, many of which landed in the upper stories of Akrotiri's abandoned homes and palaces. This caldera collapse and migration of sea water, if it was rapid enough, may itself have triggered the ensuing tidal waves (rather than the eruption itself).
5. A tsunami some 15 km long and up to 35 m high made it to Crete's northern shores within twenty minutes. It probably wiped out all of the Minoan harbors there, along with coastal villages. As it wrapped southward around

Crete, it inundated cultivated lowlands, sparing only the highest settled areas like Knossos.

6. Eruptions—and perhaps tsunamis—continued constantly over the course of approximately four days.

Scientists still debate the nuances of when the eruption occurred. While researchers hone their estimates of the year, others can show what time of year the eruption struck. Supplies in the pantries at Akrotiri were low when they were abandoned, which may indicate that produce was unharvested or still being processed in the field prior to import to the urban areas. The majority of insect eggs in the foodstuffs were unhatched; the insects in the area usually hatch in May and June. An early summer eruption also fits with the inscriptions found in Egypt.

All of the new proposed dates make it clear that the Minoan empire was not brought to an end by the eruption, as originally thought, but lived on for a century or more after. The collapse of the Minoan culture was more likely due to military or political causes. The Mycenaeans came into contact with the inhabitants of Crete and Thera toward the end of the Minoan reign (late Minoan II period, 1450–1400 B.C.), and likely played a key part in its final downfall.

Volcanoes were not the only upheaval occurring at the close of the Minoan era. In roughly 1200 B.C. Crete suffered a succession of disasters. Across the entire eastern Mediterranean, the Minoan conquerors—the Mycenaeans—experienced destruction of their own palaces and cities. Nearby, Anatolia's Hittite empire also collapsed, sending surges of refugees and armed invaders into the Aegean islands, Cyprus, Syria, and the Levant. The Kassites in Mesopotamia lost possession of their empire, and even Egypt experienced invasions leaving permanent damage to their kingdom. In the Aegean, writing appears to have completely disappeared for a time, and international trade ceased. The causes of these wide-ranging changes is not yet understood, but the result was a new dark age (Fig. 80.13).[25]

More details of Thera's last days—and the artifacts that filled its villages—will undoubtedly come to light as further investigations and advances in archeology come to the fore. But its influence throughout the Mediterranean, its elegant arts and advanced shipping, and Thera's dramatic eruption, all combine to make the Minoans one of the most important and intriguing cultures of the ancient world.

---

[25] For more on this, see *Architecture of Minoan Crete: constructing identity in the Aegean Bronze Age* by John C. McEnroe (University of Texas Press, 2010)

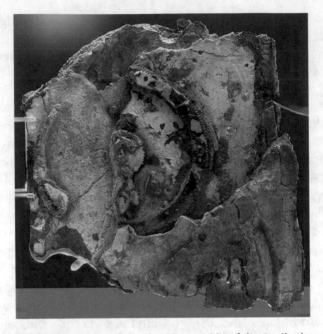

Fig. 80.13 Toothed gears emerge from the reverse side of the Antikythera Mechanism, first thought to be a rock or chunk of weathered metal. This piece, known as Fragment A, contains elements that drove the Moon calculations (the largest gear section) and part of the Saros spiral (which predicted lunar and solar eclipses). (https://commons. wikimedia.org/wiki/File:NAMA_Machine_d%27Anticyth%C3%A8re_4.jpg)

Book Club Discussion Questions

1. Trace the roles of women in the story (i.e. Mila, Katya, Kitane). What makes each one unique? Capable? What drives them?
2. Xavier considers his position to be a unique one: "The universe was a mess, and he was the only one who could put it back the way it was designed to be." Both Xavier and the Ambassador have great power available to them. How do their visions for the noble use of that power change for each of them?
3. We see Xavier descend into insanity. What are some of his symptoms? What signs do we see that Todd is following?
4. The Ambassador and Dresden are for the most part the story's villains. How is each conflicted? Do they ever present a kinder side in their actions?
5. In a novel, a major character usually has an overarching goal. Goals fall into three main categories: possession of something, revenge for something, or relief from something. Katya's initial goal for the use of time travel is complicated. How would you sum it up? What factors make her change her mind?
6. Compare the relationship of Katya and Brad to that of Todd and Kitane.
7. Katya speaks of rocking the great hourglass of time back and forth, infusing the past with future technology. If you could rock the hourglass just once and change one thing, what would it be? What might be the long-term ramifications?

M. Carroll, *Plato's Labyrinth*, Science and Fiction,
https://doi.org/10.1007/978-3-030-91709-8

8. Compare and contrast the spiritual views of Todd, Xavier, and Katya. How, if at all, do these views affect their scientific research? Their day-to-day lives? Can science and faith be comfortable bedfellows?

9. Were there factual aspects of paleontology or Minoan history that surprised you in the story?

10. For Mila van Dijk, what was the turning point, and what attitudes of the Ambassador alarmed her the most?

Further Reading

Aside from technical, peer-reviewed papers, many books used as reference for Plato's Labyrinth are available, from children's books to adult technical ones. Here is a sampling. Most of these sources are for the layperson.

Quantum Physics and Relativity

How to Build a Time Machine: the real science of time travel by Brian Clegg, St. Martin's Griffin, 2011

My First Book of Quantum Physics by Sheddad Kaid-Salah Ferron and Eduard Altarriba, Button Books, 2018

Paradox: The Nine Greatest Enigmas in Physics by Jim Al-Khalili, Broadway Books, 2012

Time Machine Tales: Science Fiction Adventures and Philosophical Puzzles of Time Travel by Paul J. Nahin (Springer 2016)

Time Traveler: a scientist's personal mission to make time travel a reality by Dr. Ronald Mallett with Bruce Henderson, Basic Books, 2009

Minoan archeology/Santorini/Ancient Rome

Architecture of Minoan Crete by John C. McEnroe (University of Texas press, 2010)

Josephus: the Jewish War edited by Gaalya Cornfeld (Zondervan Publishing, 1982). There are newer translations, but this edition has an exceptional commentary and fine background photos.

Knossos and the Herakleion Museum by Costis Davaras (Hannibal Publishing House, Athens—undated copyright)

Santorini: a guide to the island and its archaeological treasures by Christos Doumas (Ekdotike Athenon S. A. 1997)

© The Author(s), under exclusive license to Springer Nature Switzerland AG 2021
M. Carroll, *Plato's Labyrinth*, Science and Fiction,
https://doi.org/10.1007/978-3-030-91709-8

A Portable Cosmos: revealing the Antikythera mechanism, scientific wonder of the ancient world by Alexander Jones (Oxford University Press 2017)
The Society for Aegean Prehistory has an excellent website:
https://www.aegeussociety.org/en/events/excavations-and-research-ix/

Paleontology

The Dinosaurs of Waterhouse Hawkins by Barbara Kerley (Scholastic Inc, 2001)
Feathered Dinosaurs: the origin of birds by John Long and Peter Schouten (Oxford University Press 2008)
The Rise and Fall of the Dinosaurs: A New History of Their Lost World by Steve Brusatte (William Morrow Paperbacks, 2019)
The Story of Dinosaurs in 25 Discoveries: Amazing Fossils and the People Who Found Them by Donald R. Prothero (Columbia University Press, 2019)